Vaiyshnavi Perumal
Nithya Venkatachalam

Improved Scheduling Algorithm Using Dynamic Tree Construction for Wireless Sensor Networks

Anchor Academic Publishing

Perumal, Vaiyshnavi, Venkatachalam, Nithya: Improved Scheduling Algorithm Using
Dynamic Tree Construction for Wireless Sensor Networks, Hamburg, Anchor Academic
Publishing 2016

Buch-ISBN: 978-3-96067-033-9
PDF-eBook-ISBN: 978-3-96067-533-4
Druck/Herstellung: Anchor Academic Publishing, Hamburg, 2016

Bibliografische Information der Deutschen Nationalbibliothek:
Die Deutsche Nationalbibliothek verzeichnet diese Publikation in der Deutschen
Nationalbibliografie; detaillierte bibliografische Daten sind im Internet über
http://dnb.d-nb.de abrufbar.

Bibliographical Information of the German National Library:
The German National Library lists this publication in the German National Bibliography.
Detailed bibliographic data can be found at: http://dnb.d-nb.de

All rights reserved. This publication may not be reproduced, stored in a retrieval system
or transmitted, in any form or by any means, electronic, mechanical, photocopying,
recording or otherwise, without the prior permission of the publishers.

Das Werk einschließlich aller seiner Teile ist urheberrechtlich geschützt. Jede Verwertung
außerhalb der Grenzen des Urheberrechtsgesetzes ist ohne Zustimmung des Verlages
unzulässig und strafbar. Dies gilt insbesondere für Vervielfältigungen, Übersetzungen,
Mikroverfilmungen und die Einspeicherung und Bearbeitung in elektronischen Systemen.

Die Wiedergabe von Gebrauchsnamen, Handelsnamen, Warenbezeichnungen usw. in
diesem Werk berechtigt auch ohne besondere Kennzeichnung nicht zu der Annahme,
dass solche Namen im Sinne der Warenzeichen- und Markenschutz-Gesetzgebung als frei
zu betrachten wären und daher von jedermann benutzt werden dürften.

Die Informationen in diesem Werk wurden mit Sorgfalt erarbeitet. Dennoch können
Fehler nicht vollständig ausgeschlossen werden und die Diplomica Verlag GmbH, die
Autoren oder Übersetzer übernehmen keine juristische Verantwortung oder irgendeine
Haftung für evtl. verbliebene fehlerhafte Angaben und deren Folgen.

Alle Rechte vorbehalten

© Anchor Academic Publishing, Imprint der Diplomica Verlag GmbH
Hermannstal 119k, 22119 Hamburg
http://www.diplomica-verlag.de, Hamburg 2016
Printed in Germany

DEDICATED

TO

OUR BELOVED

PARENTS

ACKNOWLEDGEMENT

We extend out thanks to the almighty for grace and blessing, which enable us to complete this present endeavor in a most successful manner.

We highly obliged to my beloved Professors and friends for their valuable suggestion and kind support in all efforts. We are very thankful to our parents, siblings and relatives for their support and motivation throughout the progress of book.

We would like to express our sincere gratitude to Dr. A. Amuthan, Professor in the Department of Computer Science and Engineering, Pondicherry Engineering College, Puducherry, India and Dr. S. Lakshmana Pandian, Associate Professor in the Department of Computer Science and Engineering, Pondicherry Engineering College, Puducherry, India for their support and suggestion towards this effort.

We would like to place on record our deep sense of gratitude to Dr.R.Rathna, Assistant Professor, Department of Information Technology, Sathyabama University, and Chennai, India for her generous guidance, help and useful suggestions.

Finally, a special thanks to M.P.Suvetha, R.Vignesh Kumar, V.Karthick, K.Geetha and V.Divya Raja Krishnan for their moral support, precious love and who are always standing with us in our hard times during this work.

Thank you
Vaiyshnavi. M.P
Nithya.V

PREFACE

The Wireless Sensor Network (WSN) composed of several nodes are used for different types of monitoring applications. The objective of deploying WSN is to observe a particular site for monitoring physical parameters like temperature, light, pressure, humidity or the occurrence of a phenomenon. The Sleep/Wake up scheduling for Wireless Sensor Networks has become an essential part of its working. In this work, the details of Low Energy Adaptive Clustering Hierarchy (LEACH) which introduced the concept of clustering in sensor networks, Energy-Efficient Clustering routing algorithm based on Distance and Residual Energy for Wireless Sensor Networks (DECSA) which describes about scheduling based on distance and energy and the Energy efficient clustering algorithm for data aggregation (EECA) were discussed. The LECSA (Load and Energy Consumption based Scheduling Algorithm) has also been discussed. Based on that, in this book, the cluster head finds the nearest active node in the neighbor cluster and then it forwards its data to it. From all the cluster heads the data reaches the sink not directly, but by using a self-organized efficient routing algorithm.

TABLE OF CONTENTS

CHAPTER 1

1 INTRODUCTION 11
    1.1 Outline of the Project 19
        1.1.1 SLEEP/WAKE-UP 20
        1.1.2 DECSA 20
        1.1.3 LEACH 20
        1.1.4 EECA 21
        1.1.5 LECSA 21
        1.1.6 GSTEB 21
    1.2 Literature survey 22
    1.3 Problem Definition 25
    1.4 Objective 26
    1.5 Chapter Organization 26

CHAPTER 2

2 DESIGN AND ANALYSIS 28
    2.1 Introduction 28
    2.2 Analysis 28
        2.2.1 Project Specification 28
            2.2.1.1 Existing system 29
            2.2.1.2 Disadvantages of Existing system 29
            2.2.1.3 Proposed system 29
            2.2.1.4 Advantage of Proposed System 30
        2.2.2 Hardware and Software Requirements 30
            2.2.2.1 Hardware Requirements 30
            2.2.2.2 Software Requirements 30

2.2.2.3 Introduction to NS2 31

2.3 Design 33

   2.3.1 System Flow Diagram 34

   2.3.2 Data Flow Diagram 35

   2.3.3 UML Diagram 38

      2.3.3.1 Class Diagram 38

      2.3.3.2 Use Case Diagram 39

      2.3.3.3 Sequence Diagram 40

      2.3.3.4 Collaboration Diagram 41

      2.3.3.5 Activity Diagram 42

CHAPTER 3

3 DESIGN AND IMPLEMENTATION 43

3.1 System Architecture 43

   3.1.1 Components of GSTEB 43

      3.1.1.1 Initial Phase 43

      3.1.1.2 Tree Construction Phase 43

      3.1.1.3 Self-Organized Data Collecting &Transmitting 44

      3.1.1.4 Information Exchange 44

      3.1.1.5 Sensor Node 44

      3.1.1.6 Cluster 46

      3.1.1.7 Cluster Head 46

      3.1.1.8 Base Station 46

      3.1.1.9 End User 46

   3.1.2 Architecture of GSTEB 47

3.2 Algorithms 47

   3.2.1 Clustering 48

   3.2.2 K-Hop 48

   3.2.3 Master &Slave 49

   3.2.4 TDMA Scheduling 49

    3.2.5 LECSA 50

    3.2.6 Re-Election 51

    3.2.7 Multi-Input &Output 51

  3.3 Description Of Modules 52

    3.3.1 Cluster Formation of WSN 52

    3.3.2 LEACH 52

    3.3.3 GSTEB 52

    3.3.4 Initial Phase 53

    3.3.5 Tree Construction 53

    3.3.6 Information Exchanging 53

  3.4 Implementation 54

  3.5 Testing 55

CHAPTER 4

4 EXPERIMENTAL STUDY, RESULTS AND DISCUSSSION 57

  4.1 Description of the Experiments conducted 57

  4.2 Output with Description 57

  4.3 Experimental Results 58

    4.3.1 Throughput 59

    4.3.2 Packet Loss 60

    4.3.3 Delay Ratio 61

    4.3.4 Channel Measurement 62

    4.3.5 Protocol Frequency 63

    4.3.6 Source Frequency 64

    4.3.7 Destination Frequency 65

CHAPTER 5

5 CONCLUSION & FUTURE WORK **66**

    5.1 Summary 66

    5.2 Conclusion 67

REFERENCES **68**

A APPENDICES **70**

    A.1 Screen Shots 70

TABLE OF FIGURES

FIG No	FIGURE NAME	PAGE No
1.1	Wireless Sensor Networks	12
1.2	Sensor Node	13
1.3	DECSA Architecture	22
1.4	LEACH Architecture	23
1.5	EECA Architecture	23
2.1	Simplified User's View of N	32
2.2	System Flow Diagram	35
2.3	Level 0: Generation of Nodes	36
2.4	Level 1: Cluster Formation	36
2.5	Level 2: Initial Cluster Head Selection	37
2.6	Level 3: Scheduling and Data Transmission	37
2.7	Class Diagram	38
2.8	Use Case Diagram	39
2.9	Sequence Diagram	40
2.10	Collaboration Diagram	41
2.11	Activity Diagram	42
3.1	Sensor Node	45
3.2	GSTEB Architecture	47
4.1	Formation of cluster and Data Transfer	58
4.2	Throughput Comparison of Routing Protocols	59
4.3	Packet Loss Comparison of Routing Protocols	60
4.4	Delay Ratio Comparison of Routing Protocols	61
4.5	Channel Measurement Comparison of Routing Protocols	62
4.6	Protocol Frequency Comparison of Routing Protocols	63
4.7	Source Frequency Comparison of Routing Protocols	64
4.8	Destination Frequency Comparison of Routing Protocols	65

LIST OF SYMBOLS AND ABBREVIATIONS

ACRONYM	ABBREVIATIONS
ADC	Analog to Digital Converter
BS	Base Station
BCH	Base Station Cluster Head
CBM	Condition Based Maintenance
CH	Cluster Head
DAC	Digital to Analog Converter
DECSA	Distance Energy Cluster Structure Algorithm
DFD	Data Flow Diagram
EECA	Energy Efficient clustering Algorithm
ICH	Initial Cluster Head
ISM	Industrial, Scientific and Medical
LEACH	Low Energy Adaptive Clustering Hierarchy
LECSA	Load and Energy Consumption Based Scheduling Algorithm
NS2	Network Simulation
GSTEB	Generalized Self-Organized Tree Based Energy-Balance Routing

LIST OF TABLES

TABLE NO	TABLE NAME	PAGE NO
4.1	Configuration Parameters of the Simulation Results	57
4.2	Node IDS of Each Cluster	58

CHAPTER 1

INTRODUCTION

The sensor networks are infrastructures to collect data from the environment and the data can be used to study many problems like climate change, animal migrations, and behavior changes of buildings. The sensor nodes are deployed over a geographical area to monitor physical phenomena. The wireless sensor networks (WSN) is defined as a network of devices denoted as nodes that can sense the environment and communicate the information gathered from the monitored field through wireless data networks. The wireless sensor networks consist of hundreds of thousands of tiny, inexpensive and battery-powered wireless sensing devices which organize themselves into multi-hop radio networks.

The wireless sensor networks are a self-organizing ad hoc network with potential applications in autonomous monitoring, surveillance, military, healthcare, and security. The sensor nodes consist of three major subsystems. They are computed, communication and sensing. The computation subsystem has an embedded processor, program memory and data memory. The communication subsystem has a low power radio operating at ISM band frequency. The sensing subsystem is used to convert the external world phenomena into an equivalent electrical quantity which in turn is digitized from analog to digital converters.

The wireless sensor networks are characterized by sensing and communication coverage. The communication coverage refers to how well the sensor nodes are in communication range of each other. The sensing coverage refers to how well the terrain under monitoring is sensed by all the sensor nodes. The wireless sensor network is to determine the node density which is one of the primary challenges faced by the design of large WSN. The requirements of wireless sensor networks are fault tolerance, increased lifetime,scalability, power management, security and budget.

The wireless sensor networks are composed of tiny individual nodes that are programmed in embedded systems. They are capable of interacting with their environment through various sensors and processing information locally and communicating this information wirelessly with their neighbors.

Approved by Sathyabama University, Chennai 2015.

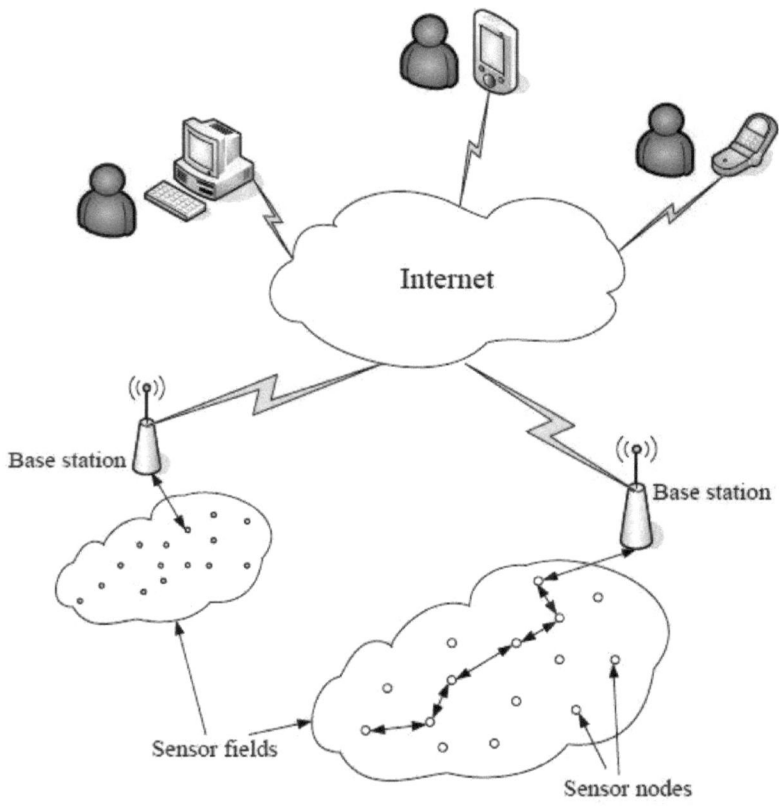

Fig: 1.1: Wireless Sensor Networks

Sensor Node

A sensor node, also known as a mote is a node in a wireless sensor network that is capable of performing some processing, gathering sensory information and communicating with other connected nodes in the network. A mote is a node, but a node cannot always be a mote.

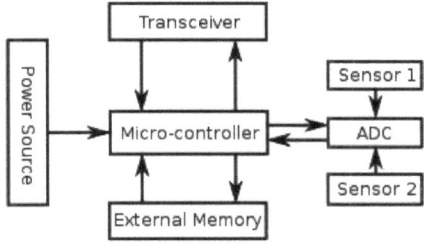

Fig: 1.2: Sensor Node

Components

The main components of a sensor node are a micro-controller, transmitter external memory, power source and one or more sensors.

Controller

The controller performs tasks, processes data and controls the functionality of other components in the sensor node. While the most common controller is a micro-controller, other alternatives that can be used as a controller are: a general purpose desktop microprocessor, digital signal processors, FPGAs and ASICs. A micro-controller is often used in many embedded systems such as sensor nodes because of its low cost, flexibility to connect to other devices, ease of programming, and low power consumption. A general purpose microprocessor generally has higher power consumption than a micro-controller; therefore it is often not considered a suitable choice for a sensor node. Digital Signal Processors may be chosen for broadband wireless communication applications, but in Wireless Sensor Networks the wireless communication is often modest: i.e., simpler, easier to process modulation and the signal processing tasks of actual sensing of data is less complicated. Therefore the advantages of DSPs are not usually of much importance to wireless sensor nodes. FPGAs can be reprogrammed and reconfigured according to requirements, but this takes more time and energy than desired.

Transceiver

Sensor nodes often make use of ISM band which gives free radio, spectrum allocation and global availability. The possible choices of wireless transmission media are Radio frequency (RF), Optical communication (Laser) and Infrared. Lasers require less energy, but need line-of-sight for communication and are sensitive to atmospheric conditions. Infrared, like lasers, needs no antenna, but it is limited in its broadcasting capacity. Radio frequency based communication is the most relevant that fits most of the WSN applications. WSNs tend to use license-free communication frequencies: 173, 433, 868, and 915 MHz; and 2.4 GHz. The functionality of both transmitter and receiver are combined into a single device known as transceivers. Transceivers often lack unique identifiers. The operational states of wireless sensor networks are transmitting, receive, idle, and sleep. Current generation transceivers have built-in state machines that perform some operations automatically.

Most transceivers operating in idle mode have a power consumption almost equal to the power consumed in receiving mode. Thus, it is better to completely shut down the transceiver rather than leave it in the idle mode when it is not transmitted or receiving. A significant amount of power is consumed when switching from sleep mode to transmit mode in order to transmit a packet.

External memory

From an energy perspective, the most relevant kinds of memory are the on-chip memory of a micro-controller and Flash memory—off-chip RAM is rarely, if ever, used. Flash memories are used due to their cost and storage capacity. Memory requirements are very much application dependent. Two categories of memory based on the purpose of storage are: user memory used for storing application related or personal data, and program memory used for programming the device. Program memory also contains identification data of the device if present.

Power source

The sensor node consumes power for sensing, communicating and data processing. More energy is required for data communication than any other process. The energy cost of transmitting 1 kB a distance of 100 meters (330 ft) is approximately the same as that used for the execution of 3 million instructions by a 100 million instructions per second/W processor. Power is stored either in batteries or capacitors. Batteries, both rechargeable and non-rechargeable, are the main source of power supply for sensor nodes. They are also classified according to electrochemical material used for the electrodes such as NiCd (nickel-cadmium), NiZn (nickel-zinc), NiMH (nickel-metal hydride), and lithium-ion. Current sensors are able to renew their energy from solar sources, temperature differences, or vibration.

Sensors

Sensors are hardware devices that produce a measurable response to a change in a physical condition like temperature or pressure. Sensors measure physical data of the parameter to be monitored. The continual analog signal produced by the sensors is digitized by an analog-to-digital converter and sent to controllers for further processing.

A sensor node should be small in size, consume extremely low energy, operate in high Volumetric densities, be autonomous and operate unattended, and be adaptive to the environment. As wireless sensor nodes are typically very small electronic devices, they can only be equipped with a limited power source of less than 0.5-2 ampere-hour and 1.2-3.7 volts.

Sensors are classified into three categories: passive, omni-directional sensors; passive, narrow-beam sensors; and active sensors. Passive sensors since the data without actually manipulating the environment by actively probing. They are self-powered; that is, energy is needed only to amplify their analog signal. Active sensors actively probe the environment, for example, a sonar or radar sensor, and they require continuous energy from a power source. Narrow-beam sensors have a well-defined notion of direction of measurement, similar to a camera. Omni-directional sensors have no notion of direction involved in their measurements.

The overall theoretical work on WSNs works with passive, omni-directional sensors. Each sensor node has a certain area of coverage for which it can reliably and accurately report the particular quantity that it is observing. Several sources of power consumption in sensors are: signal sampling and conversion of physical signals to electric ones, signal conditioning, and analog-to-digital conversion. Spatial density of sensor nodes in the field may be as high as 20 nodes per cubic meter.

Applications of Wireless Sensor Networks

Area Monitoring

Area monitoring is a common application of WSNs. In area monitoring, the WSN is deployed over a region where some phenomenon is to be monitored. A military example is the use of sensors to detect enemy intrusion; a civilian example is the Geo-fencing of gas or oil pipelines. When the sensors detect the event being monitored (heat, pressure), the event is reported to one of the base stations, which then takes appropriate action (e.g., send a message on the internet or to a satellite). Similarly, wireless sensor networks can use a range of sensors to detect the presence of vehicles ranging from motorcycles to train cars.

Air Pollution Monitoring

Wireless sensor networks have been deployed in several cities (Stockholm, London or Brisbane) to monitor the concentration of dangerous gases for citizens. These can take advantage of the ad-hoc wireless links rather than wired installations, which also make them more mobile for testing readings in different areas.

Forest Fires Detection

A network of Sensor Nodes can be installed in a forest to detect when a fire has started. The nodes can be equipped with sensors to measure temperature, humidity and gases which are produced by fires in the trees or vegetation. The earlier detection is crucial for a successful action of the firefighters; thanks to Wireless Sensor Networks, the fire brigade will be able to know when a fire is started and how it is spreading.

Greenhouse Monitoring

Wireless sensor networks are also used to control the temperature and humidity levels inside commercial greenhouses. When the temperature and humidity drops below specific levels, the greenhouse manager must be notified via e-mail or cell phone text message, or host systems can trigger misting systems, open vents, turn on fans, or control a wide variety of system responses.

Landslide Detection

A landslide detection system makes use of a wireless sensor network to detect the slightest movements of soil and changes in various parameters that may occur before or during a landslide. And through the data gathered it may be possible to know the occurrence of landslides long before it actually happens.

Machine Health Monitoring

Wireless sensor networks have been developed for machinery condition-based maintenance (CBM) as they offer significant cost savings and enable new functionalities. In wired systems, the installation of enough sensors is often limited by the cost of wiring. Previously inaccessible locations, rotating machinery, hazardous or restricted areas, and mobile assets can now be reached with wireless sensors.

Water/wastewater Monitoring

There are many opportunities for using wireless sensor networks within the water/wastewater industries. Facilities not wired for power or data transmission can be monitored using industrial wireless I/O pollution control board.

Agriculture

Wireless sensor network within the agricultural industry is increasingly common; using a wireless network frees the farmer from the maintenance of wiring in a difficult environment. Gravity feed water systems can be monitored using pressure transmitters to monitor water tank levels, pumps can be controlled using wireless I/O devices and water use can be measured and wirelessly transmitted back to a central control center for billing. Irrigation automation enables more efficient water use and reduces waste.

Structural Monitoring

Wireless sensors can be used to monitor the movement within buildings and infrastructure such as bridges, flyovers, embankments, tunnels, etc... enabling Engineering practices to monitor assets remotely without the need for costly site visits, as well as having the advantage of daily data, whereas traditionally this data was collected weekly or monthly, using physical site visits, involving either road or rail closure in some cases. It is also far more accurate than any visual inspection that would be carried out.

Natural Disaster Prevention

Wireless sensor networks can effectively act to prevent the consequences of natural disasters, like floods. Wireless nodes have successfully been deployed in rivers where changes of the water levels have to be monitored in real time.

Interior Monitoring

Observing the gas levels at vulnerable areas needs the usage of high-end, sophisticated equipment, capable to satisfy industrial regulations. Wireless internal monitoring solutions facilitate keep tabs on large areas as well as ensure the precise gas concentration degree.

Exterior Monitoring

External air quality monitoring needs the use of precise wireless sensors, rain & wind resistant solutions as well as energy reaping methods to assure extensive liberty to machines that will likely have tough access.

Environmental/Earth Monitoring

The term Environmental Sensor Networks have evolved to cover many applications of WSNs to earth science research. This includes sensing volcanoes, oceans, glaciers, forests, etc.

Air quality Monitoring

The degree of pollution in the air has to be measured frequently in order to safeguard people and the environment from any kind of damages due to air pollution.

Factors Influencing Sensor Network Design

A sensor network design is influenced by many factors, which include fault tolerance; scalability; production costs; operating environment; sensor network topology; hardware constraints; transmission media; and power consumption.

These factors are important because they serve as a guideline to design a protocol or an algorithm for sensor networks. In addition, these influencing factors can be used to compare different schemes.

Fault Tolerance

Some sensor nodes may fail or be blocked due to lack of power, physical damage or environmental interference. The failure of sensor nodes should not affect the overall task of the sensor network. This is the reliability or fault tolerance issue. Fault tolerance is the ability to sustain sensor network functionalities without any interruption due to sensor node failures. The fault tolerance level depends on the application of the sensor networks

1.1 OUTLINE OF THE PROJECT

Design of an energy efficient wireless sensor networks through clustering and scheduling based on node weighting parameter has been proposed. It is to overcome the disadvantages of LEACH, DECSA, EECA and LECSA. The LEACH (Low Energy Adaptive Clustering Hierarchy) is an application specific protocol architecture for wireless sensor networks. The DECSA (Distance and Energy Cluster Structure Algorithm) is based on the classic clustering and routing algorithm, it considers both the distance and residual energy. The EECA (Energy Efficient Clustering Algorithm) is designed to select a head node for data aggregation to reduce the energy of data transmitted. The LECSA (Load and Energy Consumption based Scheduling Algorithm) the energy efficiency of wireless sensor networks.

From LEACH, DECSA, LECSA and EECA. The GSTEB (General Self-Organized Tree-Based Energy-Balance routing protocol) in proposed has a better performance than other protocols in balancing energy consumption, thus prolonging the lifetime of WSN. By implementing it, the energy consumption of the overall network is reduced. The GSTEB which builds a routing tree using a process where, for each round, BS assigns a root node and broadcasts this selection to all sensor nodes. Subsequently, each node selects its parent by considering only itself and its neighbors' information, thus making GSTEB a dynamic protocol.

1.1.1 SLEEP/WAKE-UP

Wake-up scheduling the transceiver of sensor node have active state, sleep state and idle state. During its active state data is transferred to sink node. If the transceiver is in idle state it moves to sleep mode to save the lifetime of wireless sensor network. The energy of transceiver must be saved so it is moved to active mode at the required time. Remaining time it is moved to sleep mode. Each frame consists of sleep and wake-up mode by using the sleep/wake up protocol.

1.1.2 DECSA

In Distance Energy Cluster Structure Algorithm (DECSA) it examines both the distance and residual energy information of the nodes. DECSA protocol can be divided into initialization stage and working stage. In the initialization stage the election of cluster head is elected and coordinates with its cluster member. The cluster's head considers the node's energy consumption and communication between the node. After the election of cluster head, elect the base station cluster head based on the threshold level. In the working stage cluster head collects the data from the cluster member and transmits the data to their nearest cluster head. Then, the cluster head collects the transmitted data to the base station to balance the energy consumption and process the data transmission of the network.

1.1.3 LEACH

LEACH (Low Energy Adaptive Clustering Hierarchy) is one of the classic clustering protocols. LEACH protocol can save more energy than the plane, multi-hop routing protocols, and static network clustering algorithm. In LEACH, the nodes organize themselves into local clusters and one node acting as the cluster head. All other non-cluster head nodes transmit their data to the cluster head. The cluster head node receives data from all the cluster nodes and performs signal processing functions on the data and transmits data to the BS. The operation of LEACH is divided into rounds. Each round begins with a set-up phase when the clusters are organized, followed by a steady-state phase when data are transferred from the nodes to the cluster head on to the BS. LEACH forms clusters by using a distributed algorithm, where nodes make autonomous decisions without any centralized control.

1.1.4 EECA

Energy Efficient clustering Algorithm (EECA) is used to process the data aggregation. EECA algorithm separates the sensor network into a cluster head and its cluster member. In EECA phases can be divided into setting phase and steady pace. In setting phase cluster head allocates (TDMA) time slot to cluster members. In steady phase cluster member, send the data to the cluster head within its time slot. Then, the cluster head transmits aggregate data to sink nodes. By considering the cluster head corresponding cluster head is selected and aggregation tree is constructed to save energy.

1.1.5 LECSA

In LECSA (Load and Energy Consumption Based Scheduling) protocol is based on two functions. One is clustering and another one is scheduled. Initially cluster head is selected on highest alpha value. The Node transmits the data based on the ascending order to the cluster head. The scheduling is performed by using (TDMA) based protocol. The data can be transferred to the cluster head, then cluster head send the data to sink node. So the cluster head should have high energy transmission. In each round the cluster head can dynamically change from one node to another. So that the energy consumption of the nodes are decreased. The energy consumption is reduced.

1.1.6 GSTEB

The main task of WSN is to periodically collect information of the interested area and transmit the information to BS. A simple approach to fulfilling this task is that each sensor node transmits data directly to BS. However, when BS is located far away from the target area, the sensor nodes will die quickly due to much energy consumption. On the other hand, since the distances between each node and BS are different, direct transmission leads to unbalanced energy consumption. To solve these problems the proposed system is developed to overcome the disadvantages of existing systems.

1.2 LITERATURE SURVEY

Zhu yong, Qing peia proposed a clustering routing algorithm (DECSA), the distance energy cluster structure algorithm considering both the distance and residual energy of nodes to improve the process of cluster head selection and the process of data transmission. It reduces the adverse effect on the energy consumption of the cluster head, resulting from the nonsuniform distribution of nodes in the network and avoids direct communication between base station and cluster head, which may have low energy and far away from the base station. It resolves the limited energy of sensors in wireless sensor networks.

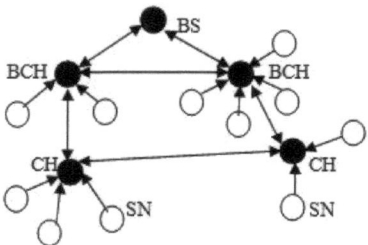

Fig: 1.3: DECSA Architecture

Wendi B. Heinzelman, Anantha P. Chandrakasan, HariBalakrishnan proposed the LEACH protocol architecture for micro sensor networks that combines the ideas of energy efficient cluster based routing and media access together with application specific data aggregation to achieve good performance in terms of system lifetime, latency and application perceived quality. It also includes a new distributed cluster formation technique that enables self organization of large no of nodes, algorithms for adapting clusters and rotating cluster head positions to evenly distribute the energy load among all the nodes and techniques to enable distributed signal processing to save communication resources.

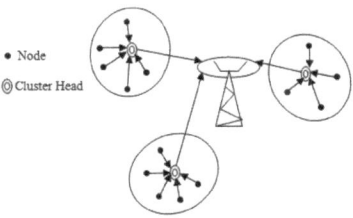

Fig: 1.4: LEACH Architecture

Sha Chao, Wang Ru-chuan, Huang Hai-ping, Sun Li-juan proposed to resolve the energy problem in wireless sensor networks. Here to select a head node for data aggregation to reduce the energy of data transmitted. The cluster head is selected by considering node's residual energy as well as the distance between this node and its neighbors. A type of data aggregation tree is constructed with the help of the clusters to reduce the amount of data transmission which effectively extends the network lifetime.

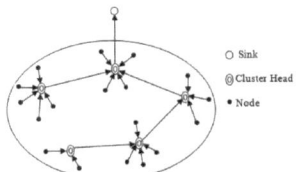

Fig: 1.5: EECA Architecture

Peter corketimwark, Raja jurdak, Wen Hu, Philip valencia, Dareenmoore proposed an application of wireless sensor network technology to the long duration and large scale environmental monitoring. The Holy Grail is a system that can be deployed and operated by domain specialists.

ViniMadan, SRN Reddy proposed the comparison between different wireless sensor motes under a number of different parameters and criteria, including processor used, expected lifetime, protocols, cost, applications.

Dilip Kumar, Trilok C. Aseri, R.B. Patel proposed the impact of heterogeneous nodes in hierarchical clusters formed in terms of their energy in wireless sensor networks. This paper introduces an energy efficient heterogeneous clustered algorithm for wireless sensor networks based on weighted election probabilities of each node to become a cluster head according to the residual energy in each node. According to this paper, the election process of cluster heads should consider the heterogeneity of the nodes, because in real time applications, not all the nodes in the field may have the same initial energy.

M. J. Handy, M. Haase, D. Timmermann proposed to reduce the overall power consumption of the wireless sensor networks. Here LEACH's stochastic cluster head selection process is extended by a deterministic component. The approach used here to increase the lifetime of the network following LEACH is the inclusion of the remaining energy of each node. This is achieved by reducing the threshold value of the competing node relative to the node's remaining energy. However, this process fails to perform after a few rounds. After a certain number of rounds the network gets stuck, though there are nodes available with enough energy to transmit data to the base station. The reason is a cluster-head threshold which is too low, because the remaining nodes have a very low energy level.

Huazhong Zhang, Peipei Chen, Shulan Gong proposed the weighted spanning tree algorithm based on LEACH (WST LEACH). Here, the selection of cluster heads are not completely random, but also considering the remaining energy, the distribution density of nodes and the distance from cluster heads to the base station. It establishes a Weighted Spanning Tree through all the cluster heads, by calculating the weight value using the aforementioned factors. Then the data is sent to the base station along this tree after being integrated. It optimizes the data transmission path, and reduces the energy consumption. Simulation results show that WST LEACH reduces the energy consumption, has higher efficiency and can extend the network lifetime as well.

Ki Young Jang, Kyung Tae Kim, Hee Yong Youn proposed a protocol to improve both the LEACH and LEACH-C algorithm. The proposed scheme works in three stages- first, electing a new cluster head based on the residual energy of each node only when the energy falls below a certain value. Second, deciding which CH to join according to the cost value determined by the signal power of the CH and the distance the CH and the respective node. Third, data transmission from the sensor nodes to the cluster head occurs only when the present condition in the context is satisfied.

Yang Tao, YalingZheng proposed an algorithm that combines the optimal number of clusters-head with energy adaptive cluster head selection algorithm. Here the initial clustering phase is divided into two phases; CH selection and cluster organization. First, each node's status is built the table at the base station according to the initial status messages (like initial energy) sent to it by the nodes and then the optimal number of CHs is selected according to the table data. In the following rounds, all the nodes in the network first record their current residual energy and calculate the value of the energy dissipated in the previous round. Then, according to the modified threshold value, each node that has elected itself a cluster-head, broadcasts an advertisement message to the rest of the nodes. According to this paper only one node will be a cluster-head successfully in a cluster of each round. This ensures that the number of clusters-heads is optimally in the whole network.

1.3 PROBLEM DEFINITION

The WSN is subjected to various resource constraints. The constraints are energy, bandwidth, memory and processing ability. Among them, energy is of prime concern, since it is severely constrained at sensor nodes and it is not feasible to either replace or recharge the batteries of sensor nodes that are often deployed in hostile environments. As a result, these constraints impose an important requirement on any QoS support mechanism in WSNs. Energy efficiency is a critical design issue in WSNs, where each sensor node relies on its limited battery power for data acquisition, processing, transmission and reception.

As the sensor nodes are typically very small and powered by irreplaceable battery, energy control becomes primary and also the most challenging problems in designing sensor networks. In WSNs, each sensor node has a different energy consumption rate due to inequality in event sensing and distance from the Base Station. This leads to energy disparity among sensor nodes in the network, which in turn shortens the lifetime of the network.

The sensor nodes operate in the three modes of sensing, computing and communications, and all of which consume energy. Of the three modes, maximum energy is expended in the communications process. The sensing unit is entrusted with the responsibility to detect the physical characteristics of the environment and has an energy consumption that varies with the hardware nature and applications. However, sensing energy represents a large percentage of the entire energy consumption within the entire WSN.

In comparison, computations energy is much more. The communication unit consists of a short-range RF circuit which performs the transmission and reception tasks. Communication energy contributes to data forwarding and it is determined by the transmission range that increases with the signal propagation in an exponential way.

1.4 OBJECTIVE

Wireless sensor network (WSN) is used to collect and send various kinds of messages to a base station (BS). Wireless sensor nodes are deployed randomly and densely in a target region, especially where the physical environment is so harsh that the macro-sensor counterparts cannot be deployed. General Self-Organized Tree-Based Energy-Balance routing protocol (GSTEB) which builds a routing tree using a process where, for each round, BS assigns a root node and broadcasts this selection to all sensor nodes. Subsequently, each node selects its parent by considering only itself and its neighbors' information, thus making GSTEB a dynamic protocol. Simulation results show that GSTEB has a better performance than other protocols in balancing energy consumption, thus prolonging the lifetime of WSN.

1.4 CHAPTER ORGANIZATION

Chapter 1 Deals with the introduction of wireless sensor networks, literature survey about the project and the origin of the project.

Chapter 2 Includes the design and analysis of the proposed method. It describes about the analysis of project specifications and hardware and software requirements. The project specifications describe about the existing system and the proposed system. The design of the project provides the various design methods to design a project.

Chapter 3 Elaborates the system design of the proposed system and various modules. The implementation and testing procedure of the project is explained.

Chapter 4 Deals with the experimental output of the project. The comparative study of the project is analyzed. The final output values are tabulated.

Chapter 5 Summarizes and concludes the project. The future work of the project is proposed.

CHAPTER 2

DESIGN AND ANALYSIS

2.1 INTRODUCTION

System analysis and design relate to shaping organizations, improving performance and achieving objectives for profitability and growth. The emphasis is on systems in action, the relationships among subsystems and their contribution to meeting a common goal. Systems, developers can generally be thought of as having two major components: Systems analysis and Systems design. System design is the process of planning a new business system or one to replace or complement an existing system. But before this planning can be done, we must thoroughly understand the old system and determine how computers can best be used to make its operation more effective. System analysis, then, is the process of gathering and interpreting facts, diagnosing problems, and using the information to recommend improvements to the system.

2.2 ANALYSIS

Analysis is the first technical step in the process of any software development. A careful analysis can help the software designer and programmer to have a better insight of the product to be created. A careless analysis can result in incomplete or dysfunctional software. To avoid such a situation, it is very important to properly identify the required software's features and create an effective design for it. It is also important to analyze and find out whether the application being developed suits the current hardware and software platform available or not. The application should be developed well within time and should meet the specified requirements.

2.2.1 Project Specification

This project is proposed to achieve energy efficiency in wireless sensor networks that helps in real – time environmental monitoring. In designing such a system, this project takes an insight into the problems related to node arrangement, data transfer and energy consumption and comes up with an algorithmic procedure.

2.2.1.1 Existing System

A main task of WSN is to periodically collect information of the interested area and transmit the information to BS. A simple approach to fulfilling this task is that each sensor node transmits data directly to BS. However, when BS is located far away from the target area, the sensor nodes will die quickly due to much energy consumption. On the other hand, since the distances are-twiners each node and BS are different, direct transmission leads to unbalanced energy consumption. To solve these problems, many protocols have been proposed. Of the protocols proposed hierarchical protocol such as LEACH.

2.2.1.2 Disadvantages of Existing System

The existing system contains various disadvantages. In LEACH protocol, the locations of nodes are not taken into an account. Due to this there is no uniform distribution among clusters. Residual Energy of nodes doesn't consider by LEACH. Due to this early death of nodes occurs. So that lifetime of the network is reduced. In LEACH there is no proper scheduling for data transmission , so packet loss is more. So its required retransmission phase.

2.2.1.3 Proposed System

A main task of WSN is to periodically collect information of the interested area and transmit the information to BS. A simple approach to fulfilling this task is that each sensor node transmits data directly to BS. However, when BS is located far away from the target area, the sensor nodes will die quickly due to much energy consumption. On the other hand, since the distances between each node and BS are different, direct transmission leads to unbalanced energy consumption. To solve these problems the proposed system is developed to overcome the disadvantages of existing systems. GSTEB is to achieve a longer network life- time for different applications.

In each round, BS assigns a root node and broadcasts its ID and its coordinates to all sensor Nodes. Then the network computes the path either by transmitting the path information from BS to sensor nodes or by having the same tree structure being dynamically and individually built by each node. For both cases, GSTEB can change the root and reconstruct the routing tree with a short delay and low energy consumption. Therefore a better balanced load is achieved compared with the protocols mentioned.

2.2.1.4 Advantages of Proposed System

In GSTEB, the load and energy considered transferring the data, and then the lifetime of the network is increased. The cluster head can dynamically change from one node to another, so that the lifetime of each node is increased. The energy consumption of the overall network is reduced. Multi input and multi output is possible in GSTEB. Master and slave concept and reelecting has been implemented.

2.2.2 Hardware and Software Requirements

This project was developed by using various hardware and software components. The hardware's and software's are essential parts to develop a project.

2.2.2.1 Hardware Requirements

Hardware is a set of physical components, which performs the functions of applying appropriate, predefined instructions. In other words, one can say that electronic and mechanical parts of computer constitute hardware.

The Hardware required for this system is as follows:

- Hard Disk: 20GB and Above
- RAM: 512MB and Above
- Processor: Pentium IV and Above

2.2.2.2 Software Requirements

The software is a set of procedures of coded information or a program which, when fed into the computer hardware enables the computer to perform the various tasks. Software is like a current in the wire, which cannot be seen, but its effect can be felt. The Software required for this system is as follows:

- C, C++ compiler for Linux
- TCL Compiler
- NS 2.35, Ubuntu 12.04

2.2.2.3 Introduction to NS2

NS (version 2) is an object-oriented, discrete event driven network simulator developed at UC Berkely written in C++ and OTcl. NS is primarily useful for simulating local and wide area networks. Although NS is fairly easy to use once you get to know the simulator, it is quite difficult for a first time user, because there are few user-friendly manuals. Even though there is a lot of documentation written by the developers which has in depth explanation of the simulator, it is written with the depth of a skilled NS user. The purpose of this project is to give a new user some basic idea of how the simulator works, how to setup simulation networks, where to look for further information about network components in simulator codes, how to create new network components, etc., mainly by giving simple examples and brief explanations based on our experiences. Although all the usage of the simulator or possible network simulation setups may not be covered in this project, the project should help a new user to get started quickly.

NS is an event driven network simulator developed at UC Berkeley that simulates a variety of IP networks. It implements network protocols such as TCP and UPD, traffic source behavior such as FTP, Telnet, Web, CBR and VBR, router queue management mechanism such as Drop Tail, RED and CBQ, routing algorithms such as Dijkstra, and more. NS also implements multicasting and some of the MAC layer protocols for LAN simulations. The NS project is now a part of the VINT project that develops tools for simulation results display, analysis and converters that convert network topologies generated by well-known generators to NS formats. Currently, NS (version 2) written in C++ and OTcl (Tcl script language with Object-oriented extensions developed at MIT) is available. This document talks briefly about the basic structure of NS, and explains in detail how to use NS mostly by giving examples. Most of the figures that are used in describing the NS basic structure and network components are from the 5th VINT/NS Simulator Tutorial/Workshop slides and the NS Manual (formerly called "NS Notes and Documentation"), modified a little bit as needed. For more information about NS and the related tools, visit the VINT project home page.

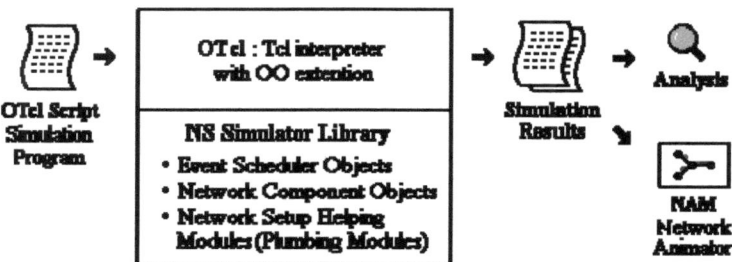

Figure 2.1. Simplified User's View of NS

As shown in Figure2.1, in a simplified user's view, NS is Object-oriented Tcl (OTcl) script interpreter that has a simulation event scheduler and network component object libraries, and network setup (plumbing) module libraries (actually, plumbing modules are implemented as member functions of the base simulator object). In other words, to use NS, you program in OTcl script language.

To setup and run a simulation network, a user should write an OTcl script that initiates an event scheduler, sets up the network topology using the network objects and the plumbing functions in the library, and tells traffic sources when to start and stop transmitting packets through the event scheduler. The term "plumbing" is used for a network setup, because setting up a network is plumbing possible data paths among network objects by setting the "neighbor" pointer of an object to the address of an appropriate object. When a user wants to make a new network object, he or she can easily make an object either by writing a new object or by making a compound object from the object library, and plumb the data path through the object. This may sound like complicated job, but the plumbing OTcl modules actually make the job very easy. The power of NS comes from this plumbing.

Another major component of NS beside network objects is the event scheduler. An event in NS is a packet ID that is unique for a packet with scheduled time and the pointer to an object that handles the event. In NS, an event scheduler keeps track of simulation time and fires all the events in the event queue scheduled for the current time by invoking appropriate network components, which usually are the ones who issued the events, and let them do the appropriate action associated with packet pointed by the event.

Network components communicate with one another, passing packets, however, this does not consume actual simulation time. All the network components that need to spend some simulation time handling a packet (i.e. Need a delay) use the event scheduler by issuing an event for the packet and waiting for the event to be fired to itself before doing further action handling the packet. After all, network configurations, scheduling and post-simulation procedure specifications are done, the only thing left is to run the simulation. This is done by $ns run.

C++ Compiler

GCC is a portable compiler--it runs on most platforms available today, and can produce output for many types of processors. In addition to the processors used in personal computers, it also supports microcontrollers, DSPs and 64-bit CPUs. GCC is not only a native compiler--it can also cross-compile any program, producing executable files for a different system from the one used by GCC itself. This allows software to be compiled for embedded systems which are not capable of running a compiler. GCC is written in C with a strong focus on portability, and can compile itself, so it can be adapted to new systems easily. GCC has multiple language front ends, for parsing different languages. Programs in each language can be compiled, or cross-compiled, for any architecture..

2.3 DESIGN

Based on the user requirements and the detailed analysis of the existing system, the new system must be designed. This is the phase of system designing. It is the most crucial phase in the development of a system. The logical system design arrived at as a result of systems analysis is converted into a physical system design.

Normally, the design proceeds in two stages

    Preliminary or General Design

    Structured or Detailed Design

Preliminary or General Design

In the preliminary or general design, the features of the new system are specified. The costs of implementing these features and the benefits to be derived are estimated. If the project is still considered to be feasible, we move to the detailed design stage.

Structured or Detailed Design

In the detailed design stage, computer oriented work begins in earnest. At this stage, the design of the system becomes more structured. Structure design is a blue print of a computer system solution to a given problem having the same components and interrelationships among the same components as the original problem. Input, output, databases, forms, codification schemes and processing specifications are drawn up in detail.

2.3.1 System Flow Diagram

A system flow diagram simply shows the breakdown of a task or system into all of the necessary steps. Each step is represented by a symbol and connecting lines show the step-by-step progression through the task. There must be a clear start, series of steps, direction of flow and a clear finish or finish point.

The system flow diagram contains the start and end functions. The various steps of the project can be explained in figure 2.2.

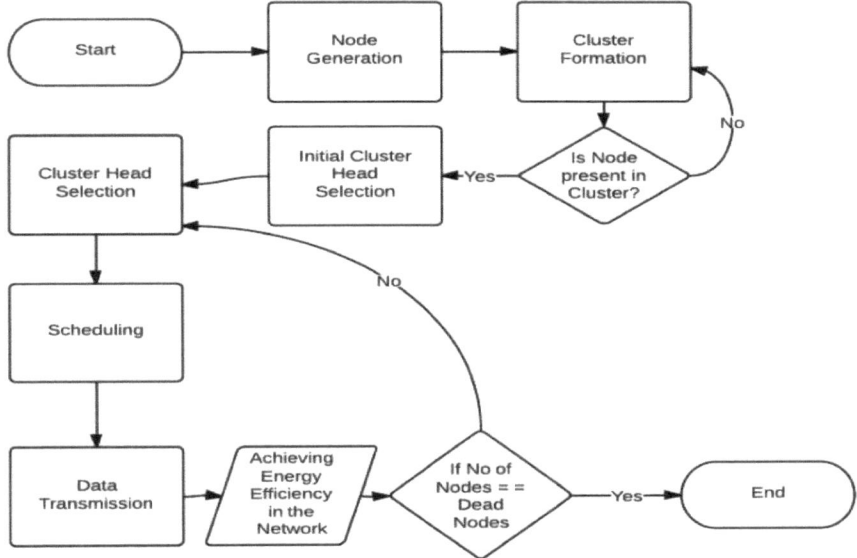

Fig: 2.2: System Flow Diagram

2.3.2 Data Flow Diagram

The Data Flow Diagram (DFD) is a graphical representation of the flow of data through an information system. It enables you to represent the processes in your information system from the viewpoint of data. The DFD lets you visualize how the system operates, what the system accomplishes and how it will be implemented, when it is refined with further specification. Data flow diagrams are used by systems analysts to design information-processing systems, but also as a way to model whole organizations.

Level 0: Generation of Nodes

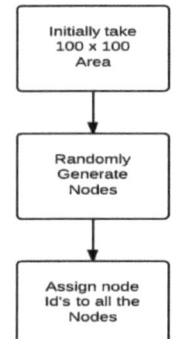

Fig: 2.3: Level 0 - Generation of Nodes

Level 1: Cluster Formation

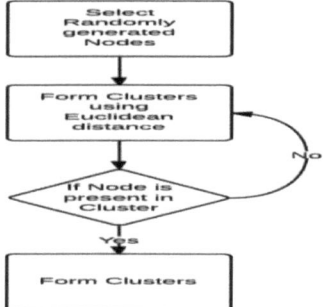

Fig: 2.4: Level 1 – Cluster Formation

Level 2: Initial Cluster Head Selection

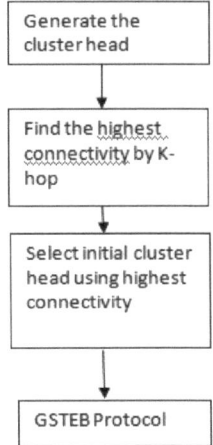

Fig: 2.5: Level 2 – Initial Cluster Head Selection

Level 4: Scheduling and Data Transmission

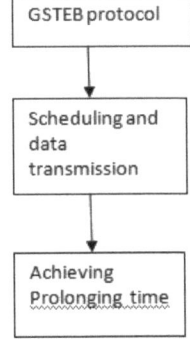

Fig: 2.6: Level 3 - Scheduling and Data Transmission

2.3.3 UML Diagrams

The UML represents a collection of best engineering practices that have proven successful in the modeling of large and complex systems. The UML is very important parts of developing object oriented software and the software development process. The UML uses mostly graphical notations to express the design of software projects. Using the UML helps project teams communicate, explore potential designs, and validate the architectural design of the software.

2.3.3.1 Class Diagram

Class diagram describes the types of objects in the system and the various kinds of static relationships that exist among them. It shows the properties and operations of a class and the constraints that apply to the way objects are connected. A class box has three parts, Name of the class, Attributes of the class, and Operations of the class.

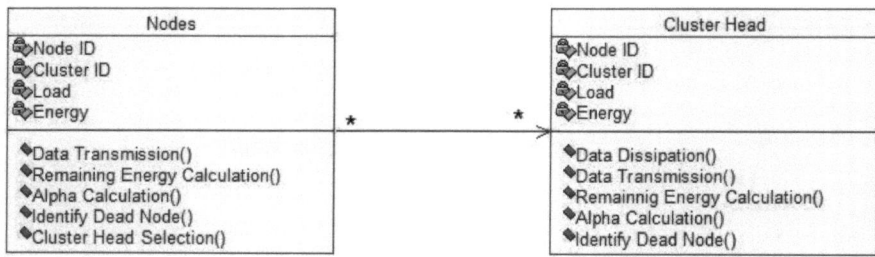

Fig: 2.7: Class Diagram

This project contains two classes, one is nodes and another one is cluster head. The node's class contains few attributes and some operations. The attributes are node ID, cluster ID, load and energy. The operations are data transmission, remaining energy calculation, and alpha calculation, identify dead nodes and cluster head selection. The cluster head class contains few attributes and some operations. The attributes are node ID, cluster ID, load and energy. The operations are data dissipation, data transmission, remaining energy calculation, alpha calculation, and identify the dead node.

2.3.3.2 Use Case Diagram

Use cases serve as a technique for capturing the functional requirements of a system. It describes the typical interactions between the users of a system and the system itself, providing a narrative of how a system is used. A use case consists of a set of one or more scenarios tied together by a common user goal. A scenario is a sequence of steps describing an interaction between a user and a system, some scenarios describe successful interaction, and others describe failure or errors.

In this project three actors involved to transfer the data. They are nodes, parent node, leaf node, base station. The nodes actor performs various use case functions.. The parent node actor performs signal processing, tree construction, monitor channel, neighbor information. The leaf node actor performs sending ack, broadcast efficient. The base station node actor performs the data transfer function, self organization, energy propagation effect, load balancing.

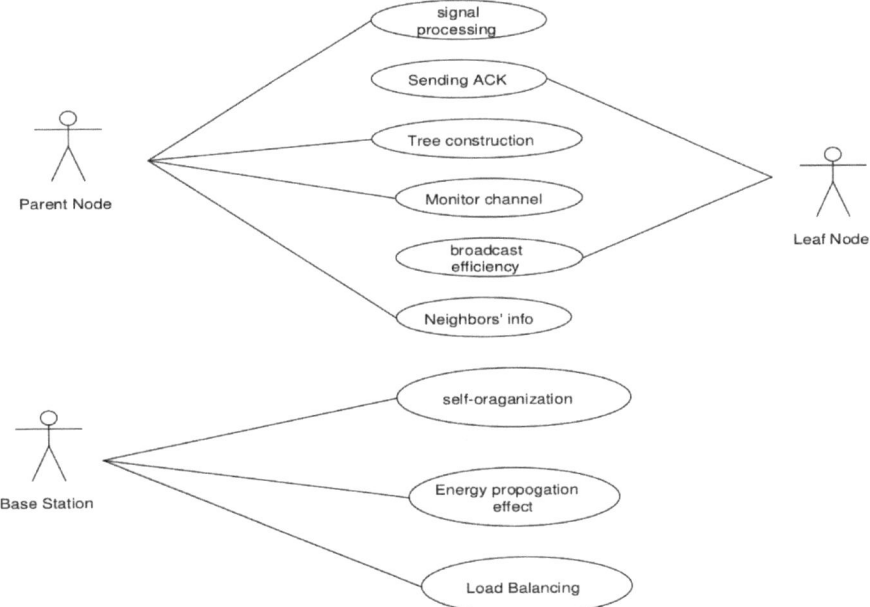

Fig: 2.8: Use Case Diagram

39

2.3.3.3 Sequence Diagram

The properties represent structural features of a class and consist of attributes and associations. It captures the behavior of a single scenario in a use case. It shows a number of example objects and messages that are passed between those objects within the use case. The columns of the diagram represent each object involved in the use case. The lifetime of an object progresses from the top of the diagram to the bottom.

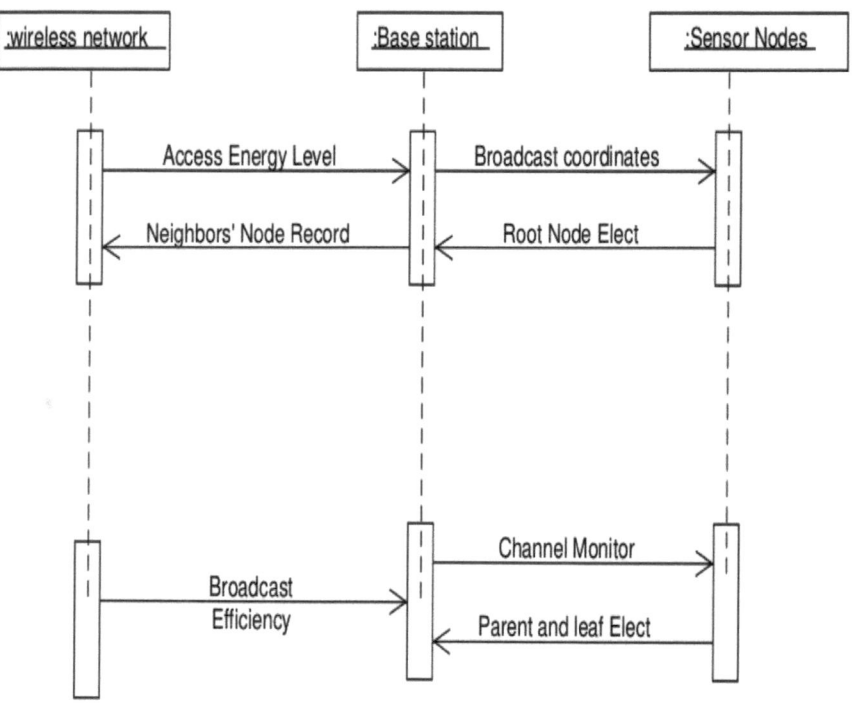

Fig: 2.9: Sequence Diagram for Nodes

2.3.3.4 Collaboration Diagram

Dynamic behavior of objects can, in addition to sequence diagrams, also be represented by collaboration diagrams. The transformation from a sequence diagram into a collaboration diagram is a bi-directional function. The difference between sequence diagrams and collaboration diagrams is that collaboration diagrams emphasize more the structure than the sequence of interactions. Within sequence diagrams the order of interactions is established by vertical positioning, whereas in collaboration diagrams the sequence is given by numbering the interactions.

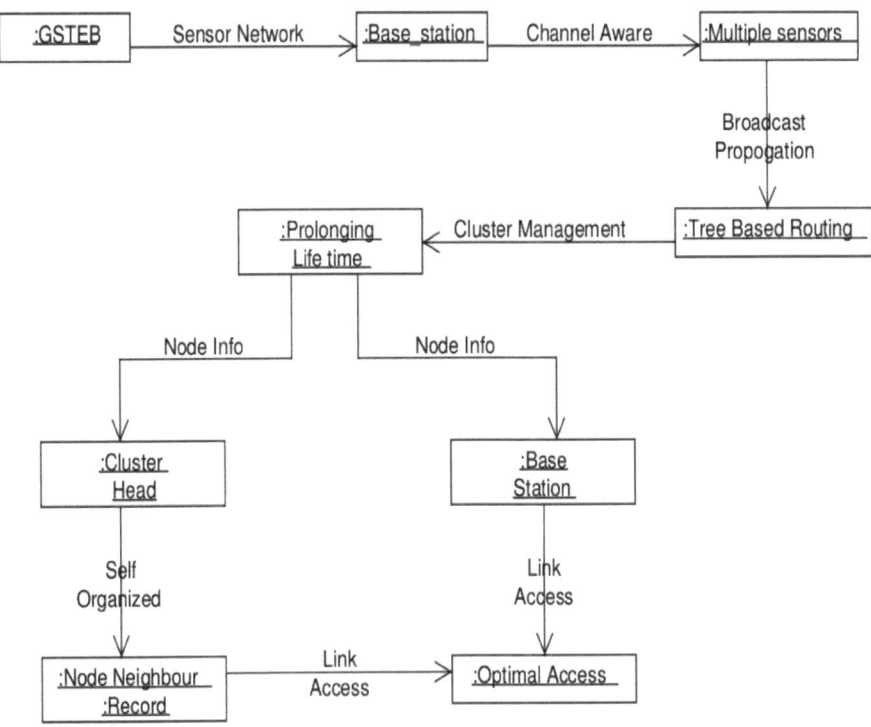

Fig: 2.10: Collaboration diagram for Nodes

2.3.3.5 Activity Diagram

It serves as a technique to describe procedural logic, business process logic, and workflow. It is similar to a flowchart except that it can also show a parallel behavior. It states the essential sequencing rules to follow, thereby allowing concurrent algorithms to be used. Consequently, an activity diagram allows whoever is doing the process to choose the order in which to do certain things.

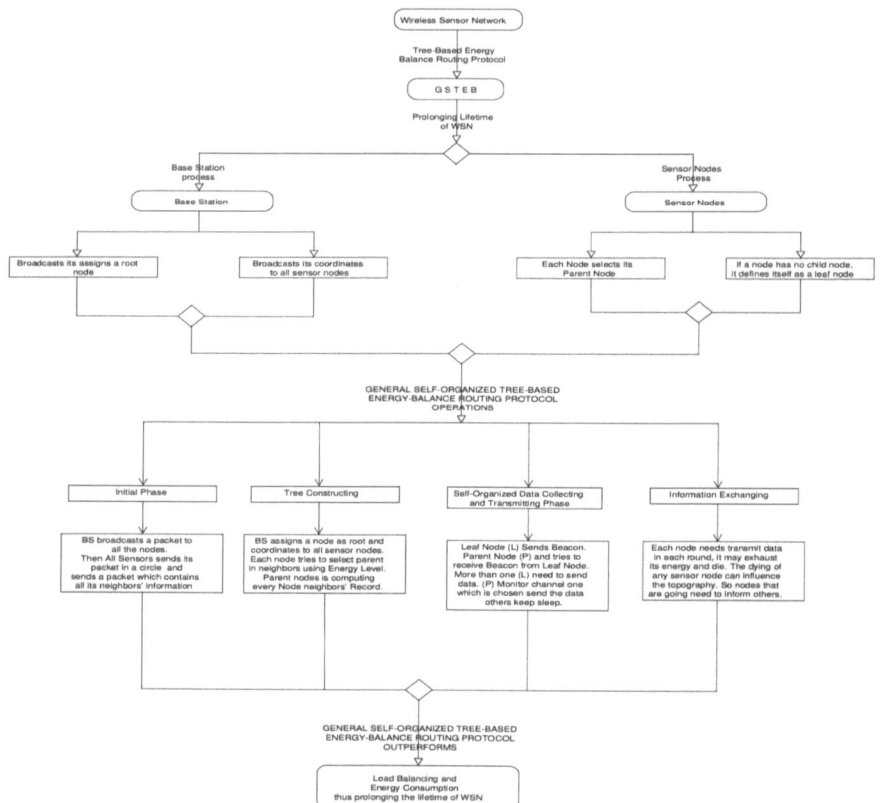

Fig: 2.11: Activity Diagram

CHAPTER 3

DESIGN AND IMPLEMENTATION

3.1 SYSTEM ARCHITECTURE

The System Architecture consists of various components. The various modules are described to implement the GSTEB. The GSTEB is proposed to achieve the better performance than other protocols in balancing energy consumption, thus prolonging the lifetime of WSN energy efficiency of the network.

3.1.1 Components of GSTEB

GSTEB is to achieve a longer network lifetime for different applications. The operation of GSTEB is divided into Initial Phase, Tree Constructing, Self-Organized Data Collecting and Transmitting Phase, Information Exchanging, are Sensor Nodes, Clusters, Cluster Heads, Base Station and End Users

3.1.1.1 Initial Phase

In this phase, the distributed cluster formation occurs. The initial Cluster Head (CH) circulates token among its neighbors. Those nodes which are in its range are detected by this method and clusters are formed using K-means clustering. The data travels from the node where it is first sensed to the sink.

If the time consumption for this transfer is more, then the data from the CH travels through the intermediate CHs and reach the sink.BS broadcasts a packet to all the nodes. Then All Sensors send its packet in a circle and sends a packet which contains all its neighbors' information.

3.1.1.2 Tree Construction Phase

In this phase, the tree is constructed based on the neighbor node information. Every parent node maintains a record of its child nodes. If a node does not have a child node then it defines itself as leaf node.BS assigns a node as root and coordinates to all sensor nodes.

Each node tries to select a parent in neighbors using Energy Level. Parent nodes are computed every Node neighbors' Record.

3.1.1.3 Self-Organized Data Collecting and Transmitting Phase

Leaf Node (L) Sends Beacon. Parent Node (P) and tries to receive Beacon from Leaf Node. More than one (L) need to send data. (P) Monitor channel one which is chosen send the data others keep sleeping.

3.1.1.4 Information Exchange

In this phase, all the parent nodes transmit their data to the root node. If a child node is about to die, it discharges its' data to next immediate active node. In short time, the tree has to be reconstructed using this information. While forming the clusters threshold value is used to have a check over the cluster size.

When the CH receives (Join-message) sent by the ordinary node, it will compare the size of the cluster with the threshold to accept new member and update the count of cluster nodes. If the request is rejected and that node already has a CH, then the clustering process ceases. Otherwise, it finds another appropriate cluster to join. During every round, the clustering takes place and the details of the re- elected CHs are informed to the sink. Each node needs to transmit data in each round; it may exhaust its energy and die. The dying of any sensor node can influence the topography. So nodes that are going need to inform others.

3.1.1.5 Sensor Node

A sensor node is the core component of a WSN. Sensor nodes can take on multiple roles in a network, such as simple sensing, data storage, routing and data processing. The Sensor node consists of various components.

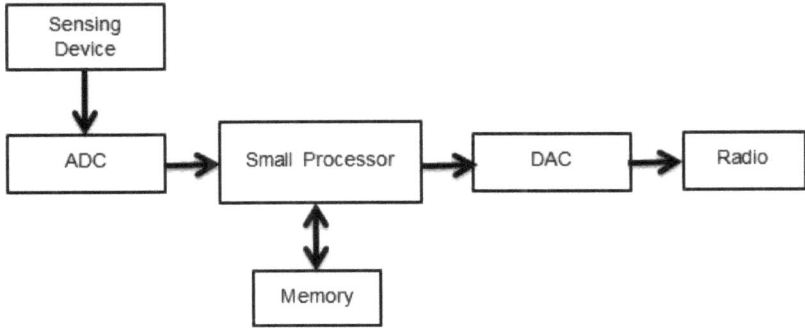

Fig: 3.1: Sensor Node Architecture

Sensing Device

It is the detection head of the WSN and is not only involved in radiation detection, but also provides inputs pertaining to density of radiation in a specific area and hence requirements for sleep and wake up process.

ADC

It provides the conversion of data from its analog perceivable form to digital format in order to be processed by the system.

Small processor

This is the heart of our WSN and provides the computation of the necessary data in order to decide the data to be transferred and the processor fetches the information of the system from the memory to decide the data transfer course.

Memory

It contains the data related to cluster nodes and related information such as sink node allotment, nearest node information and energy values.

DAC

The transferring data provided by the processor in digital format are converted here to provide input to the radio for sending it to the destination node.

Radio

This is the transmission channel of the WSN. The data are transferred to the designated node based on the information provided by the processor.

3.1.1.6 Cluster

The dense nature of these networks requires the need for them to be broken down into clusters to simplify tasks such a communication.

3.1.1.7 Cluster Head

Cluster Heads are the organizational leader of a cluster. They often are required to organize activities in the cluster. These tasks include, but are not limited to data-aggregation and organizations the communication schedule of a cluster.

3.1.1.8 Base Station

The Base Station is used to communicate with all the nodes. The Base station is used to communicate between end users and sensor nodes. It used to transfer data from one node to another. The base station is involved in the data transmission. It transmits data to the end users.

3.1.1.9 End Users

The data in a sensor network can be used for a wide-range of applications. Therefore, a particular application may make use of the network data over the internet, using a PDA, or even a desktop computer. In a queried sensor network (where the required data is gathered from a query sent through the network). This query is generated by the end user. The end user consists of Internet links, PDA terminal, user terminal.

3.1.2 Architecture of GSTEB

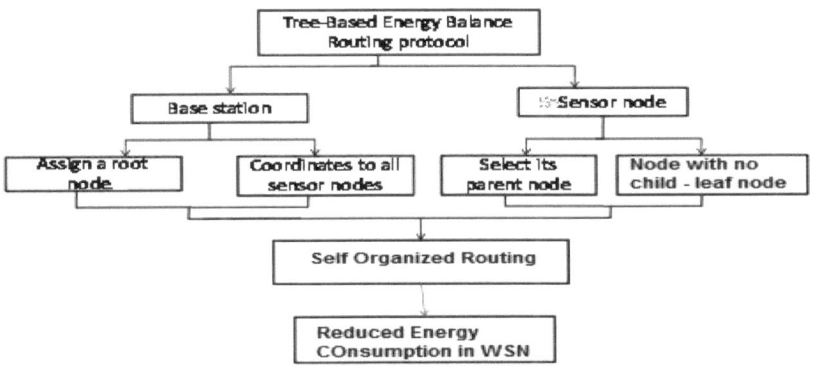

Fig: 3.2: GSTEB Architecture

The main aim of GSTEB is to achieve a longer network lifetime for different applications. In each round, BS assigns a root node and broadcasts its ID and its coordinates to all sensor nodes. Then the network computes the path either by transmitting the path information from BS to sensor nodes or by having the same tree structure being dynamically and individually built by each node. For both cases, GSTEB can change the root and reconstruct the routing tree with a short delay and low energy consumption. Therefore a better balanced load is achieved compared with the protocol. The operation of GSTEB is divided into Initial Phase, Tree Constructing Phase, Self-Organized Data Collecting and Transmitting Phase, and Information Exchanging Phase.

3.2 ALGORITHMS

Based on the clustering and scheduling, the GSTEB is implemented. There are clustering, k-hop, master and slave, Re-election, multi Input and output and scheduling. The scheduling is based on the TDMA Scheduling. The scheduling is used to avoid collision in the data transfer.

3.2.1 Clustering

Clustering is the procedure used to group a number of similar things in order to achieve a common goal. This goal may be either a complex computation, which requires more computing power than every single node can offer, or collecting data (sensor data for example) from a wide geographic area. T

he concept of similarity may be specified using many different parameters related to the specific problem we are trying to solve, such as proximity to the rest of the nodes, computing and processing ability, mobility in the area of the network, the energy requirement for the operation of the node or any combination of the above parameters. Also with this grouping of the network, we achieve a level of role assignment inside the clusters and the network in order to reduce the volume of traffic generated, to control and synchronize the networks function.

With clustering each node in the network assumes a certain role and along with that role some privileges and some obligations in order to achieve the common goal. These roles can be easily distinguished, and are: Leader or cluster head is the node that is responsible for each one of the clusters. Simple nodes are those that have no special role in the clusters. Gateway nodes are located on the outskirts of the clusters and connect them to the rest of the network.

3.2.2 K-Hop

The purpose is to minimize the number of clusters formed in the network and in this way obtain dominating sets of smaller sizes. Clusters in the K- CONID approach are formed by a cluster head and all nodes that are at distance at most k-hops from the cluster head At the beginning of the algorithm, a node starts a flooding process in which a clustering request is sent to all other nodes. In the Highest-degree heuristic, node degree only measures connectivity for 1-hop clusters. K-CONID generalizes connectivity for a k-hop neighborhood. Thus, when $k = 1$ connectivity is the same as node degree. Each node in the network is assigned a pair $did = (d, ID)$. d is a node's connectivity and ID is the node's identifier. A node is selected as a cluster head if it has the highest connectivity. In case of equal connectivity, a node has a cluster head priority if it has lowest ID. The basic idea is that every node broadcasts its clustering, decision once all its k-hop neighbors with larger clustered priority have done.

3.2.3 Master and slave:

The purpose of this scheme is to minimize the transmission energy consumption summed by all master- slave pairs and to serve as many slaves as possible in order to operate the network with longer lifetime and better performance. Two schemes, single-phase clustering and double-phase clustering, are proposed. In single-phase clustering, initially every master node will page slave nodes with the allowed maximum energy. For each slave that receives one or multiple paging signal, it always sends an acknowledgment message back to the master from which it receives the strongest paging signal. Since a master node can serve only a limited number of slaves, it first allocates channels for slaves that only receive a single paging signal from itself. If any free channels remain, other slave nodes, which receive more than one paging signal, are allocated channels in the order of the power level of the paging signal received from the master node. For those slave nodes, which do not receive a channel from a master in the channel allocation phase, are dropped in the further communication phase.

3.2.4 TDMA Scheduling

TDMA scheduling can be defined as the process of allocating time slots to the nodes or links between each pair of neighboring nodes, to ensure collision free channel access. The TDMA scheduling is divided into two types. They are Broadcast Scheduling and Link Scheduling.

• *Broadcast Scheduling:* The stations themselves are scheduled. The transmission of a station must be received collision-free by all its one-hop neighbors

• *Link Scheduling:* The links between stations are scheduled. The transmission of a station must be received collision-free by one particular neighbor.

The objective of TDMA Scheduling is to get rid of *primary* and *secondary conflict*.

• *Primary conflict:* Occurs when one node transmits and receives at the same time slot or receives more than one transmission destined to it at the same time slot.

• *Secondary conflict:* Occurs when an intended receiver of a particular transmission is also within the transmission range of other transmission intended for other nodes.

3.2.5 LECSA (LOAD AND ENERGY CONSUMPTION BASED SCHEDULING ALGORITHM

The LECSA is a scheduling algorithm meant for environmental monitoring applications. According to that algorithm, all the nodes are randomly distributed in a given area. A shortest path tree is selected. Based on the position of nodes and according to this tree structure, the nodes are clustered . The nodes which are within a distance of 10m are grouped together as clusters. If a node is satisfying this condition for two clusters, then that node is added to a cluster which has least number of nodes. Initially for the first cycle all the nodes in the network generate a random number between 0 and 1 for every node in the network as their α parameter. The nodes in the cluster compare their α values and the one which has higher α value is selected as the initial Cluster Head (ICH). Then every node works continuously.

For the second cycle, the α value is calculated based on the residual node energy and the data packets it has. Less energy indicates that the node is soon going to die. If the node has more data packets, then it should be given a chance to send the data as soon as possible. So every node has to be weighted based on the above two factors. Hence the alpha value is calculated by the following equation .

$$\alpha = K * Eit + \frac{1}{Lit}$$

Where Eit denotes residual node power (remaining power) of the node i at time t and Lit denote the load of the node i (number of data packets it has at the instant t). Then by comparing this α parameter, the node which has the maximum α value is selected as the Cluster Head (CH).

All the nodes in the cluster send their data according to the schedule created based on their α value. The one which is having very low α value occupies the initial slot of the schedule for becoming active. Then, according to the increasing order of the α value the nodes are given slots to transfer their data to the Cluster Head. Since the CH is having the highest α value, it has more residual node power and it remains active for the entire cycle. Every node has the data aggregation capability. In this design any node becomes the CH based on its residual energy and load. The CH has to aggregate the data and forward. At the next higher level, all the CHs have to forward their data to the next CH in line to reach the sink node. So the CH of a node forwards the data to a node in the neighbor cluster

3.2.6 Re-election

Clustering schemes may cause the cluster structure to be completely rebuilt over the whole network when some local events take place, e.g. the movement or "die" of a mobile node, resulting in some cluster head re-election (re-clustering). This is called the ripple effect of re-clustering. In other words, the ripple effect of re-clustering indicates that the re-election of one cluster head may affect the structure of many clusters and arouse the cluster head re-election over the net- work. Thus, the ripple effect of re-clustering may greatly affect the performance of upper-layer protocols. In addition, most schemes separate the clustering into two phases, cluster formation and cluster maintenance, and assume that mobile nodes keep static when cluster formation is in progress. This is because of the initial cluster formation of these schemes, a mobile node can decide to become a cluster head only after it exchanges some specific information with its neighbors and assures that it holds some specific attribute in its neighborhood. With a frozen period of motion, each mobile node can obtain accurate information from neighboring nodes, and the initial cluster structure can be formed with some specific characteristics. However, this assumption may not be applicable in an actual scenario where mobile nodes may move randomly all the time.

3.2.7 Multi Input and Output

Two clusters may deploy the same frequency or code set if they are not neigh- boring clusters. Also, a cluster can better coordinate its transmission events with the help of a special mobile node, such as a cluster head, residing in it. This can save much resources used for retransmission resulting from reduced transmission collision. The second benefit is in routing, because the set of cluster heads and cluster gateways can normally form a virtual backbone for inter-cluster routing, and thus the generation and spreading of routing information can be restricted in this set of nodes. Last, a cluster structure makes an adhoc network appear smaller and more stable in the view of each mobile terminal. When a mobile node changes its attaching cluster, only mobile nodes residing in the corresponding clusters need to update the information. Thus, local changes need not be seen and updated by the entire network, and information processed and stored by each mobile node are greatly reduced.

3.3 DESCRIPTION OF MODULES

The GSTEB is composed of various modules. Each module describes the different operations to achieve the energy efficiency. The modules are based on the functions of the GSTEB. They are

- Cluster formation
- LEACH
- GSTEB operation
- Initial Phase
- Tree construction phase
- Information exchange

3.3.1 Cluster formation of WSN

WSN is to regularly collect information from the sensor node and transmit it to cluster head. The cluster heads are selected by using highest node weighting parameter. Cluster head is organized by the entire sensor node and keeps track of the information of all the sensor nodes. Once the Sensor node is deployed, they will keep operating until there is a discharge.

3.3.2 LEACH (LEACH Low Energy Adaptive Clustering Hierarchy)

WSN is considered a dynamic clustering method. All the nodes can transmit enough power to reach to the base station and the nodes use power control. The LEACH This network is made up of nodes, which are called as cluster heads. The work of the cluster head is to gather the data from their nearest nodes and transfer it on to the sink node. The LEACH has two phases. They are set-up phase, steady-state phase. In set-up phase cluster heads are chosen. In steady-state the cluster heads are maintained. Every node could transmit data into the corresponding time-slot By single hop communication.

3.3.3 Generalized Self-Organized Tree Based Energy-Balance [GSTEB]

In each round the sink node assigns a root node and coordinates its sensor nodes. Root node sends the time slot message to all the sensor nodes. These sensor nodes send its message in a round which contains the entire neighbor's information. If the sensor node doesn't receive the message means not in the range, so the sensor node moves to sleep state. Therefore a better balanced load is achieved compared with the protocols mentioned.

3.3.4 Initial Phase

When initial Phase begins token analysis in random allocation. By using Distributed Self – Organization balanced clustering algorithm can divide into three phases. They are cluster head selecting phase, cluster building phase, cycle phase. In the cluster head selecting phase cluster head node broadcasts (Head-Message) through the token analysis. Once all the nodes received this message data packet start traveling form cluster head to cluster members.

In the cluster selecting phase can be divided into two phases. There are set-up phase and steady pace. In the set-up phase cluster are organized by cluster head. Cluster in the K-hop approach forms cluster head. At the beginning time slot the k-hop neighbors to declare itself as cluster head and ask them to join the cluster. The neighbors in k-hop to calculate the respective weight, and then the node with the highest weight will become the cluster head. In the steady phase cluster head transfer the data to the base station.

In the cluster building phase cluster head send the data to the base station if it's too long to reach the base station, that time cluster building phase generate. By using this phase cluster header to send the data to the nearest active cluster head from their data send to the base station. By this communication cost is reduced. In the last cycle phase is reelecting process.

3.3.5 Tree Construction

The tree is constructed based on the neighbor's information. Each node selects its parent node by the sensor node information and its energy. The parent node is computed every node's neighbor record. If a node has no child node, it defines itself as a leaf node. Parent node directly communicates with the root node and manages the entire child node.

3.3.6 Information Exchanging

While exchanging information in each round, Parent node has to transmit the data to the root node. If any child node discharges its energy and die. The dying child node should inform to its neighbor node before discharging. Thus, the tree is reconstructed within short time to save the energy by using this tree based routing load and energy consumption of network. GSTEB is based on the connectivity density and the distance from the base station.

GSTEB sets the threshold size for all the cluster head. The cluster head manages cluster and forwards data, so it consumes energy faster than the other node. The number of cluster nodes cannot exceed the threshold size. To avoid forming large cluster, which will cause extra overhead and reduce network lifetimes. When the cluster head node receives (Join-message) sent by the ordinary node, it will compare the size of the cluster with the threshold to accept new member and update the count of cluster nodes. If the size is smaller than the threshold, it will reject the request. If the rejected request has a cluster head already, the clustering process ceases.

Otherwise, it finds another appropriate cluster to join. Multi input and multi output is possible in GSTEB. By using this save the resources used for transmission and reduced transmission collision. One round has been completed next cycle phase is reelecting the cluster. The reelecting cluster is based on the cluster head gather the weight of all the member nodes, and then selects the node with highest weight as the next cluster head node. The reelecting information updates in base station. Master and slave concept has been implemented by GSTEB. Master mean it will manage all the slave nodes. The slave receives and send message to the master. A master node sends one paging to the slave node. If the slave node receives the paging mean it will send back to master node. Since the master node manage only a limited number of slave node. By using this master and slave approach to minimize the transmission energy consumption and better performance. Some time malicious gives misbehaved in communication signal. So the node won't give proper signal strength. Cluster based certificate revocation for enlisting and removing the certificates of nodes that have been detected to launch attacks on the neighborhood.

3.4 IMPLEMENTATION

Using NS2, the GSTEB can be implemented. Initially the nodes can be randomly placed using the random function. There are hundred random values are assumed to the x-axis and y-axis to plot the nodes in 100m × 100m area field. The sink node is placed in the middle of the area field. All the values are assumed to the all the nodes. Using the K-hop in NS2, the clusters are formed. The clusters are differentiated using different color codes in NS2.

For the first round, a node is selected as a cluster head if it has the highest connectivity. And the cluster head can receive the data from the all the nodes in the cluster and cluster head can transfer the data to the base station. If the base station is too near it will send directly or if it's too long it will send the data to the nearest cluster head from there to base station. Reelecting is done. Again cluster head can be selected based on the highest connectivity. Same as round one, the data transmission and data dissipation are done. The node loses its energy then the node is considered as a dead node. The dead node is not involved in the data transmission and data dissipation. After many rounds all the node reaches the dead node condition. Then the system has stopped its functions.

3.5 TESTING

Before implementing the new system into operation, a test run of the system is done for removing the bugs, if any. It is an important phase of a successful system. After codifying the whole programs of the system, a test plan should be developed and run on a given set of test data. The output of the test run should match the expected results. Sometimes, testing is considered a part of the implementation process.

Using the test data following test run is carried out:

Program Test
System Test
Unit Test
Acceptance Test
Integration Test

Program Test

When the programs have been coded, compiled and brought to working conditions, they must be individually tested with the prepared test data. Any undesirable happening must be noted and debugged (error corrections).

System Test

After carrying out the test program for each of the programs of the system and errors removed, then system test is done. At this stage the test is done on actual data. The complete system is executed on the actual data. At each stage of the execution, the results or output of the system is analyzed. During the result analysis, it may be found that the outputs are not matching the expected output of the system. In such case, the errors in the particular programs are identified and are fixed and further tested for the expected output.

Unit Test

In computer programming, unit testing is a method of testing that verifies the individual units of source code are working properly. A unit is the smallest testable part of an application. In procedural programming a unit may be an individual program, function, procedure, etc., while in object-oriented programming, smallest unit is a method, which may belong to a base/super class, abstract class or derived/child class.

Acceptance Test

Acceptance testing checks the system against the "requirements". It is similar to system testing in that the whole system is checked, but important difference is a change in focus: systems testing checks that the system that was specified has been delivered. Acceptance testing checks that the system delivers what was required.

Integration Test

Integration testing' (sometimes called integration and testing, abbreviated I&T) is the phase of software testing in which individual software modules are combined and tested as a group. It follows unit testing and precedes testing. Integration testing takes as its input modules that have been unit tested, groups them in larger aggregates, applies tests defined in an integration test plan to those aggregates. Only it is ensured that the system is running error-free, the actual data will be supplied to the system for real time implementation.

CHAPTER 4

EXPERIMENTAL STUDY, RESULTS AND DISCUSSION

4.1 DESCRIPTION OF THE EXPERIMENTS CONDUCTED

The simulation environment with parameters taken for GSTEB is shown down. The described GSTEB algorithms in chapter 3 have been carried out in NS2. The configuration is shown in Table 4.1.

Table 4.1: Configuration parameters of the simulation model

Number of nodes	50
Area size	50*50
Simulation time	20,40,60,80sec
Packet size	80 bytes

The details of simulation parameters are as follows: In an area of 50*50 m2 sensor fields. 40 sensor nodes deployed. The packet size is 80bytes.

4.2 OUTPUT WITH DESCRIPTION

The results obtained from the simulation are compared with LEACH and GSTEB. The node generated using the random function. Based on the random values of x-axis and y-axis, the nodes can be placed in different areas. There are 40 nodes can be placed in the 50 × 50. The nodes are grouped together and form a cluster. The first node is selected and calculates the K-hop distance. It starts generating the clusters by the highest connectivity. And the highest connectivity value of each cluster is selected as Initial Cluster Head. The data can be transferred from each node to cluster head. The cluster head transfers the data to the sink node. Same as previous procedure, the data can be transferred from each node to cluster head and cluster head to sink node.

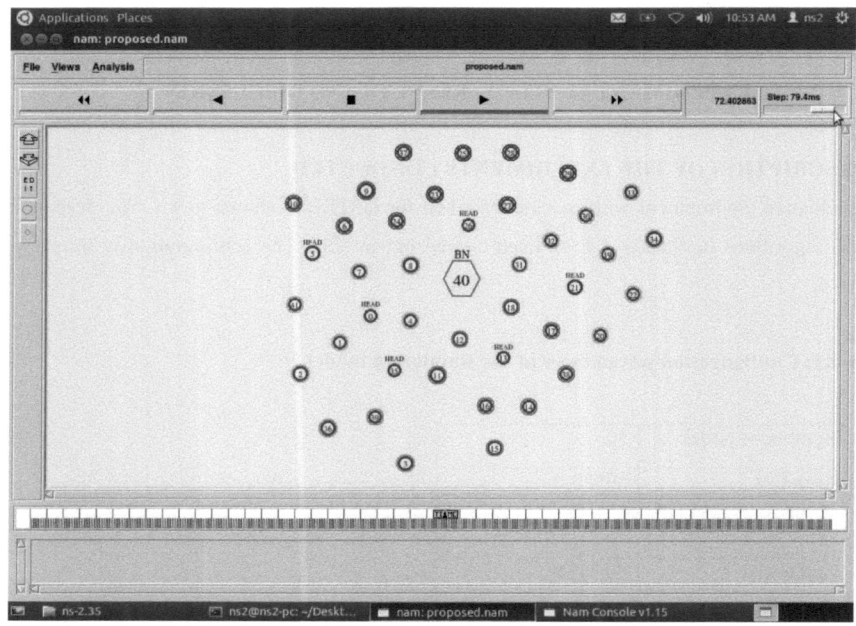

Fig: 4.1: Formation of cluster and data transfer

4.3 EXPERIMENTAL RESULTS (GRAPHS, TABLES)

Using ns2 Simulation, the clusters are formed using randomly generated nodes. The node can be identified by node IDs. The clusters are differentiated using different colors. The configuration is shown in Table 4.2.

Table 4.2: Node IDs of Each Cluster

Cluster 1	Cluster 2	Cluster 3	Cluster 4	Cluster 5	Cluster 6
0(CH)	35(CH)	13(CH)	21(CH)	26(CH)	5(CH)
1	39	12	31	23	10
2	36	18	32	37	9
41	3	11	30	25	6
7		17	34	28	24
8		38	19	27	
4		16	20		
		14	22		
		15	33		

4.3.1 Throughput

It is the number of packets successfully received by the receiver. The graphical representation of throughput comparison is shown in the figure 4.2.

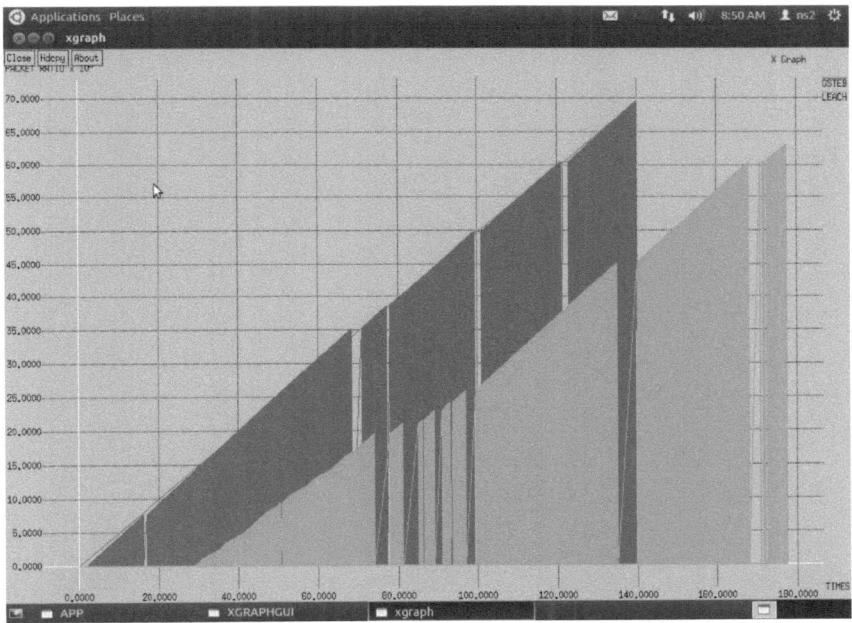

Figure 4.2: Throughput Comparison of Routing Protocols

The graph shoes that the proposed protocol is better than the existing protocols such as LEACH and GSTEB.

4.3.2 Packet Loss

It is number of packet loss during the data transmission.The graphical representation of the packet loss comparison is shown in the figure 4.3.

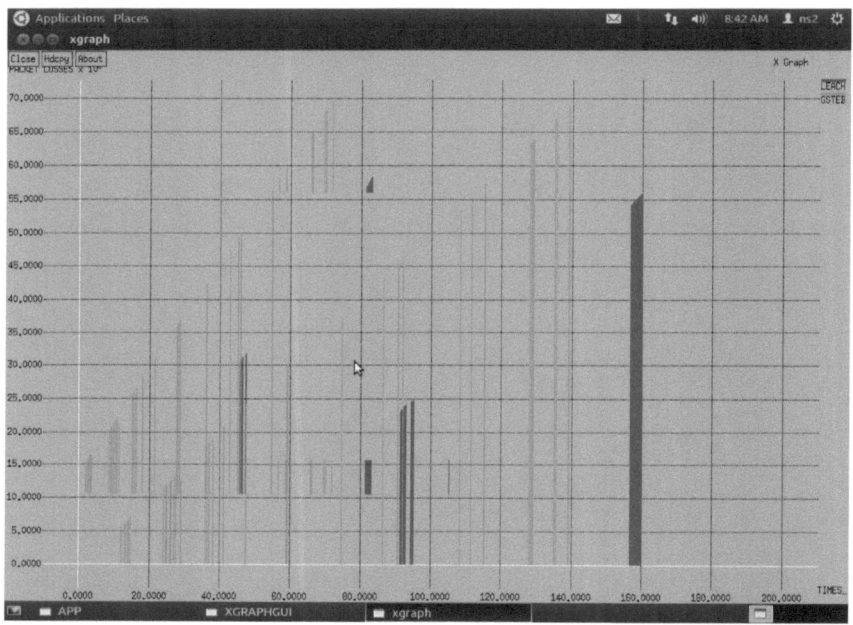

Figure 4.3: Packet Loss Comparison of Routing Protocols

The graph shoes that the proposed protocol is better than the existing protocols such as LEACH and GSTEB. The packet loss is lesser when compared with existing algorithms such as LEACH and GSTEB.

4.3.3 Delay Ratio

The graphical representation of the packet loss comparison is shown in the figure 4.4.

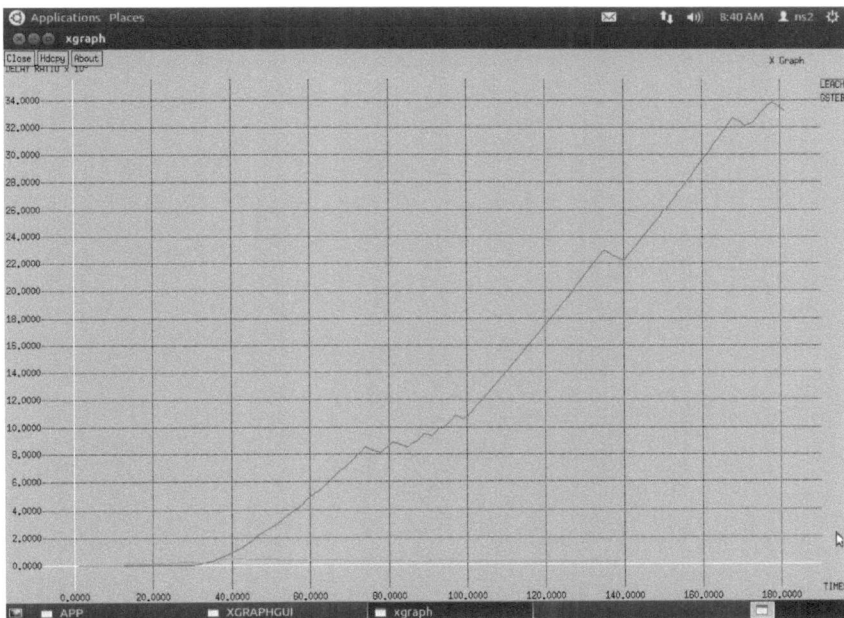

Figure 4.4: Delay Ratio Comparison of Routing Protocols

The graph shoes that the proposed protocol is better than the existing protocols such as LEACH and GSTEB. The Delay Ratio is lesser when compared with existing algorithms such as LEACH and GSTEB.

4.3.4 Channel Measurement

The graphical representation of the packet loss comparison is shown in the figure 4.5.

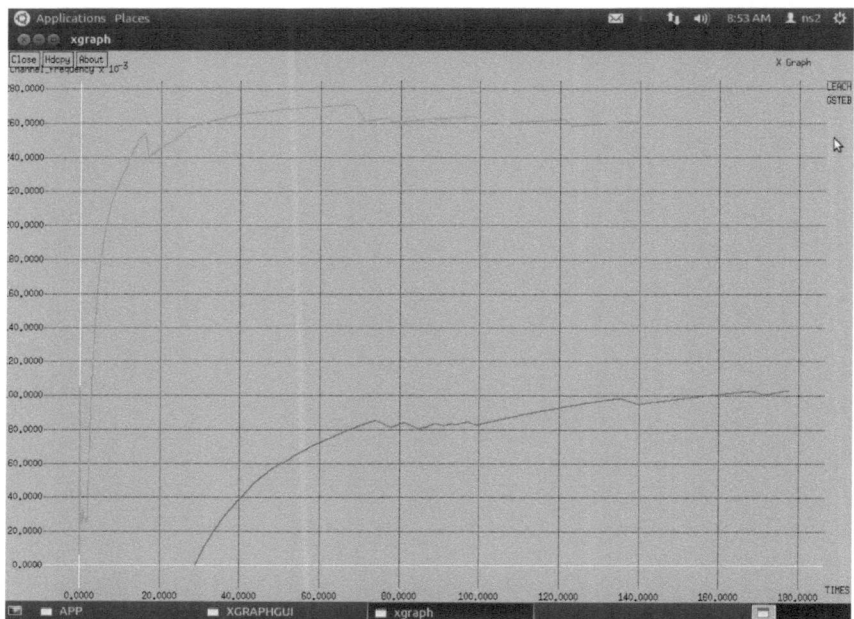

Figure 4.5: Channel Measurement Comparison of Routing Protocols

The graph shoes that the proposed protocol is better than the existing protocols such as LEACH and GSTEB. The Channel Measurement is greater when compared with existing algorithms such as LEACH and GSTEB.

4.3.5 Protocol Frequency

The graphical representation of the packet loss comparison is shown in the figure 4.6.

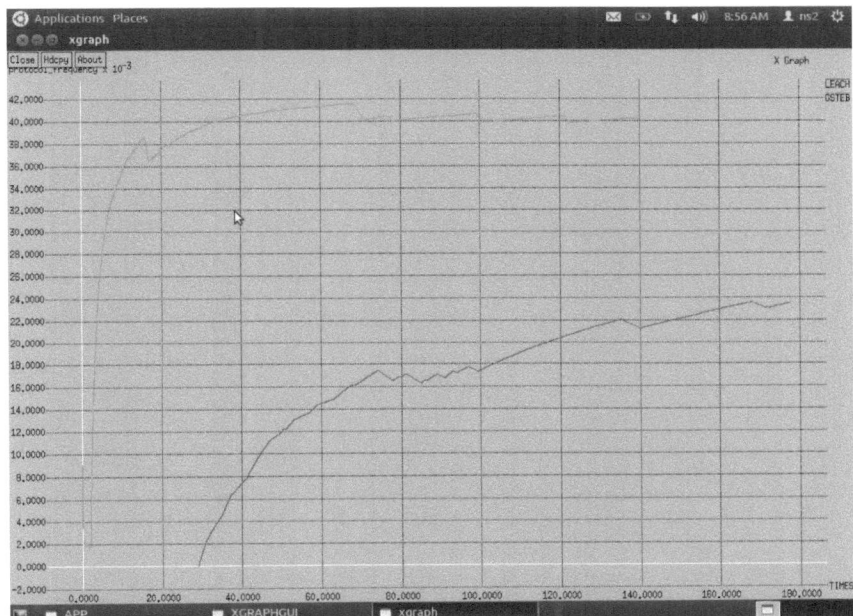

Figure 4.6: Protocol Frequency Comparison of Routing Protocols

The graph shoes that the proposed protocol is better than the existing protocols such as LEACH and GSTEB. The Protocol frequency is greater when compared with existing algorithms such as LEACH and GSTEB.

4.3.6 Source Frequency

The graphical representation of the packet loss comparison is shown in the figure 4.7.

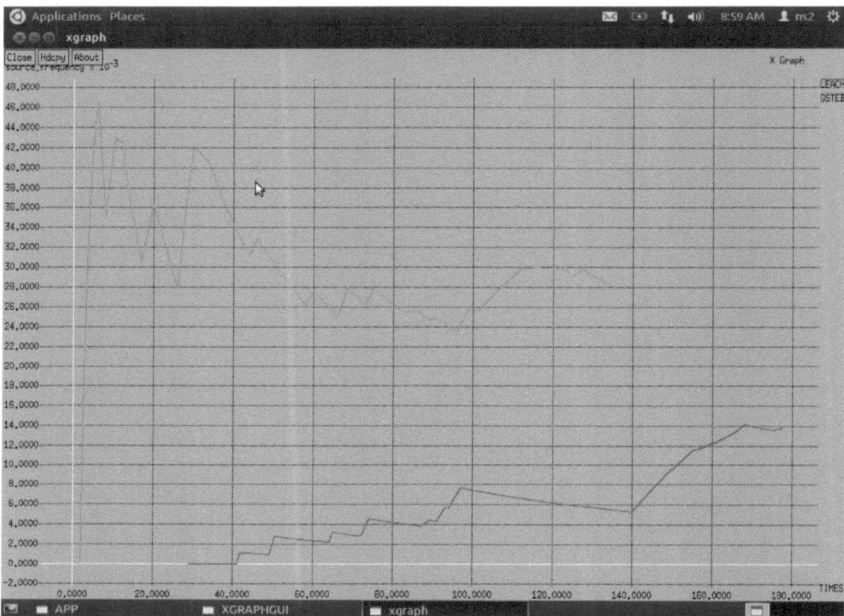

Figure 4.7: Source Frequency Comparison of Routing Protocols

The graph shoes that the proposed protocol is better than the existing protocols such as LEACH and GSTEB. The Source frequency is greater when compared with existing algorithms such as LEACH and GSTEB.

4.3.7 Destination Frequency

The graphical representation of the packet loss comparison is shown in the figure 4.8.

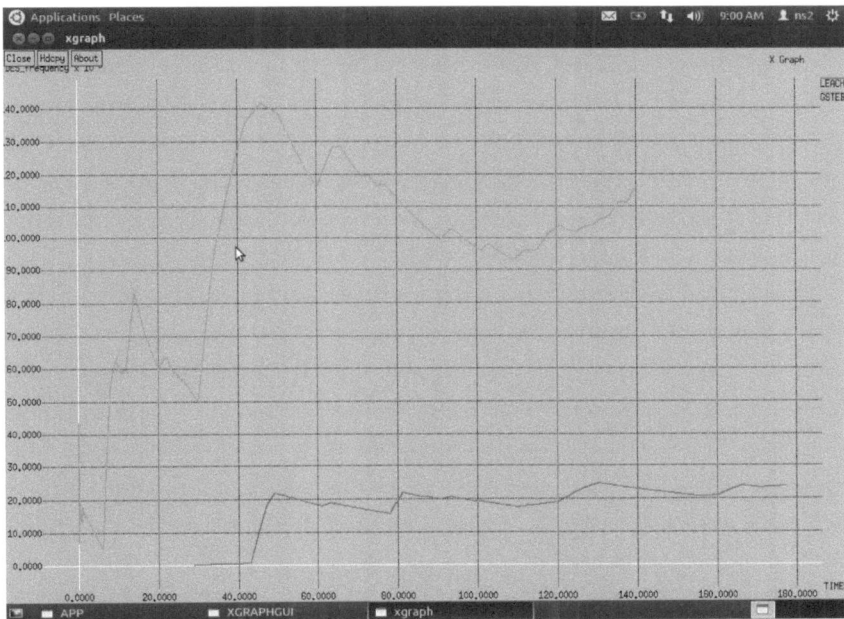

Figure 4.8: Destination Frequency Comparison of Routing Protocols

The graph shoes that the proposed protocol is better than the existing protocols such as LEACH and GSTEB. The Source frequency is greater when compared with existing algorithms such as LEACH and GSTEB.

CHAPTER 5

CONCLUSION AND FUTURE WORK

5.1 SUMMARY

In this work, GSTEB is implemented. Two definitions of network lifetime and two extreme cases of data fusion are discussed. It offers another simple approach to balance the network load. In fact, it is difficult to distribute the load evenly on all nodes in such a case. Even though GSTEB needs BS to compute the topography, which leads to an increase in energy waste and a longer delay, this kind of energy waste and longer delay are acceptable when compared to the energy consumption and the time delay for data transmitting.

Simulation results show that when lifetime is defined as the time from the start of the network operation to the death of the first node in the network, GSTEB prolongs the lifetime of the network to a greater extent when compared with LEACH. In each round the sink node assigns a root node and coordinates its sensor nodes. Root node sends the time slot message to all the sensor nodes. These sensor nodes send their message in a round which contains the entire neighbor's information. If the sensor node doesn't receive the message it means that it is not in the range. so the sensor node moves to sleep state. Therefore a better balanced load is achieved compared with the protocols mentioned.

GSTEB sets the threshold size for all the cluster head. The cluster head manages cluster and forwards data, so it consumes energy faster than the other node. The number of cluster nodes cannot exceed the threshold size. To avoid forming large cluster, which will cause extra overhead and reduce network lifetimes. When the cluster head node receives (Join-message) sent by the ordinary node, it will compare the size of the cluster with the threshold to accept new member and update the count of cluster nodes. If the size is smaller than the threshold, it will reject the request. If the rejected request has a cluster head already, the clustering process ceases. Otherwise, it finds another appropriate cluster to join.

Multi input and multi output is possible in GSTEB. By using this save the resources used for transmission and reduce transmission collision. After the completion of one round the next cycle phase is reelects the cluster Head. The reelecting cluster is based on the cluster head together with the weight of all the member nodes, and then selects the node with highest weight as the next cluster head node. The reelecting information updates in base station.

Master and slave concept has been implemented by GSTEB. Master means it will manage all the slave nodes. The slave receives and send message to the master. A master node sends one paging to the slave node. If the slave node receives the paging mean it will send back to master node, Since the master node manage only a limited number of slave node. master and slave approach can be used to minimize the transmission energy consumption and better performance.

5.2 CONCLUSION

Zhao Han, Jie Wu, *Member, IEEE*, Jie Zhang, Liefeng Liu, and Kaiyun Tian," General Self-Organized Tree-Based Energy-Balance routing protocol" in the IEEE TRANSACTIONS ON NUCLEAR SCIENCE, VOL. 61, NO. 2, APRIL 2014.

In this work, GSTEB protocol is used for energy efficient routing. Two definitions of network lifetime and two extreme cases of data fusion are discussed. The NS2 simulations show that when the data collected by sensors is strongly correlative, GSTEB works best. GSTEB is a self-organized protocol, it only consumes a small amount of energy . All the leaf nodes can transmit data in the same TDMA time slot so that the transmitting delay is short. GSTEB offers another simple approach to balancing the network load. In fact, it is difficult to distribute the load evenly on all nodes, if existing protocols are used.

REFERENCES

[1] Zhao Han, jie Wu, "A General Self-Organized Tree-Based Energy-Balance Routing Protocol for Wireless Sensor Network"- Nuclear Science, IEEE Transactions on - ISSN : 0018-949910 Vol.61 10. April 2014.

[2] K. Akkaya and M. Younis, "A survey of routing protocols in wireless sensor networks," *Elsevier Ad Hoc Network J.*, vol. 3/3, pp. 325–349,2005.

[3] I. F. Akyildiz *et al.*, "Wireless sensor networks: A survey," *Computer Netw.*, vol. 38, pp. 393–422, Mar. 2002.

[4] W. R. Heinzelman, A. Chandrakasan, and H. Balakrishnan, "Energy efficient communication protocols for wireless microsensor networks," in *Proc. 33rd Hawaii Int. Conf. System Sci.*, Jan. 2000, pp. 3005–3014.

[5] W. B. Heinzelman, A. Chandrakasan, and H. Balakrishanan, "An application-specific protocol architechture for wireless microsensor networks," *IEEE Trans.Wireless Commun*, vol. 1, no. 4, pp. 660–670,Oct. 2002.

[6] S. Lindsey and C. Raghavendra, "Pegasis: Power-efficient gathering in sensor information systems," in *Proc. IEEE Aerospace Conf.*, 2002, vol. 3, pp. 1125–1130.

[7] W. Liang and Y. Liu, "Online data gathering for maximizing network lifetime in sensor networks," *IEEE Trans Mobile Computing*, vol. 6, no. 1, pp. 2–11, 2007.

[8] R. Rathna, A. Sivasubramanian, Vinoth Kumar, "Load and Energy Consumption based Scheduling Algorithm for Wireless Sensor Networks (LECSA)" -Indian Journal of Computer Science and Engineering (IJCSE) - ISSN: 0976-5166 Vol. 4 No.6 Dec 2013-Jan 2014.

[9] SHA Chao, WANG Ru-chuan, HUANG Hai-ping, SUN Li-juan, "Energy efficient clustering algorithm for data aggregation in wireless sensor networks". The Journal of China Universities of Posts and Telecommunications, December 2010, 17(Suppl. 2): 104–109.

[10] Sohrabi *et al.*, "Protocols for self-organization of a wireless sensor network," *IEEE Personal Commun.*, vol. 7, no. 5, pp. 16–27, Oct. 2000.

[11] S. S. Satapathy and N. Sarma, "TREEPSI: Tree based energy efficient protocol for sensor information," in *Proc. IFIP Int. Conf.*, Apr. 2006, pp. 11–13.

[12] H. O. Tan and I. Korpeoglu, "Power efficient data gathering and aggregation in wireless sensor networks," *SIGMOD Rec.*, vol. 32, no. 4, pp.66–71, 2003.

[13] Xiaohua Xu, Xiang-Yang Li, Min Song "Efficient Aggregation Scheduling in Multihop Wireless Sensor Networks with SINR Constraints in the IEEE Transactions on Mobile Computing, Vol. 12, No. 12, December 2013.

[14] O. Younis and S. Fahmy, "HEED: A hybrid, energy-efficient, distributed clustering approach for ad hoc sensor networks," *IEEE Trans Mobile Computing*, vol. 3, no. 4, pp. 660–669, 2004.

APPENDICES

A.1 SCREEN SHOTS

1. The Random Displacements of Sensor Nodes

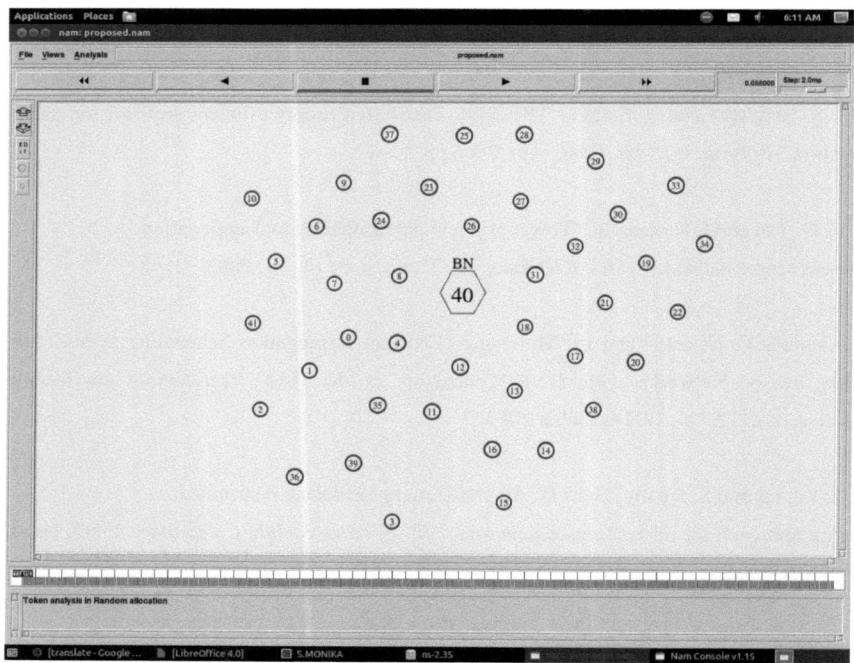

2. Cluster Formation

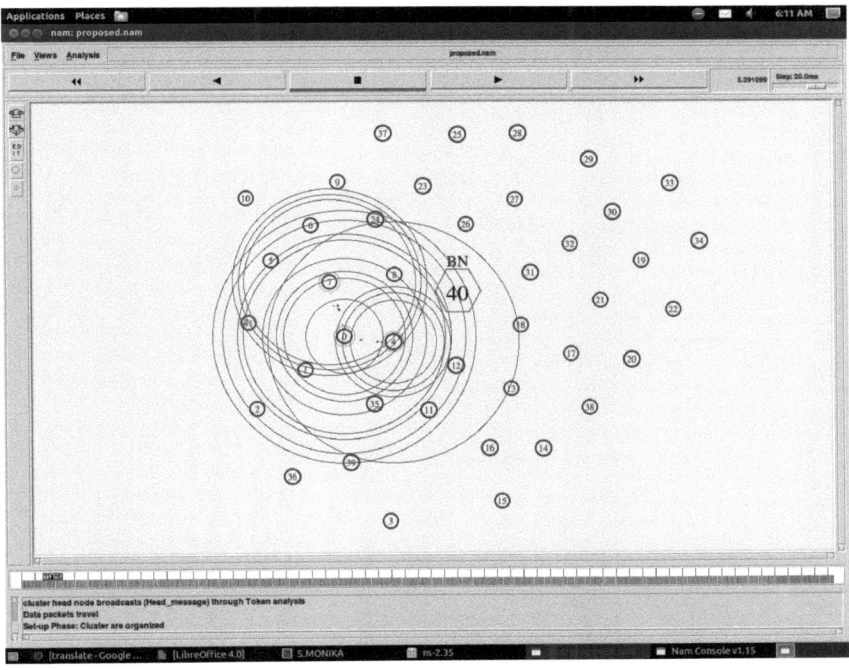

3. Initial Cluster Head Selection

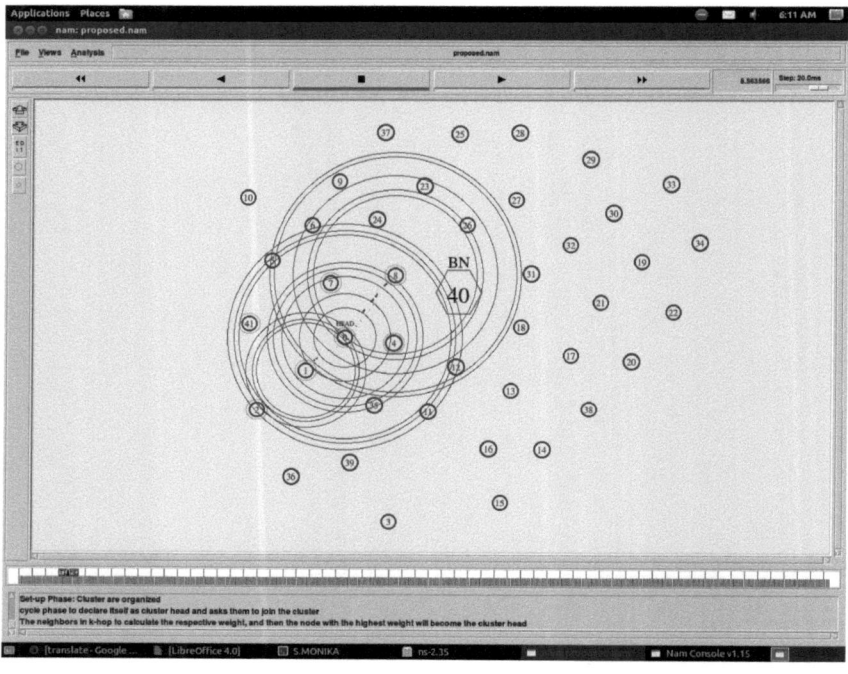

4. Cluster Head Selection and Data Transmission

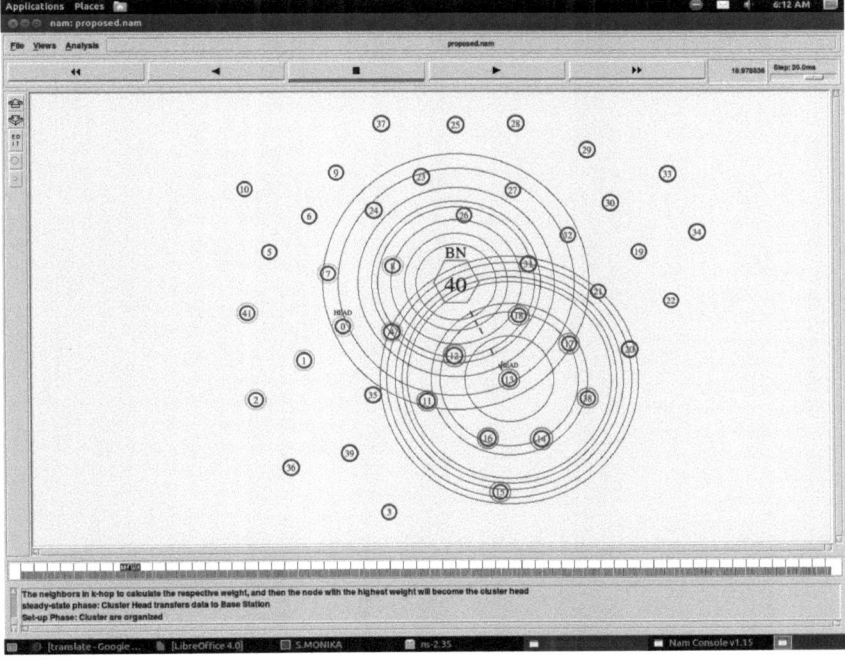

5. Tree Structure Data Transmission

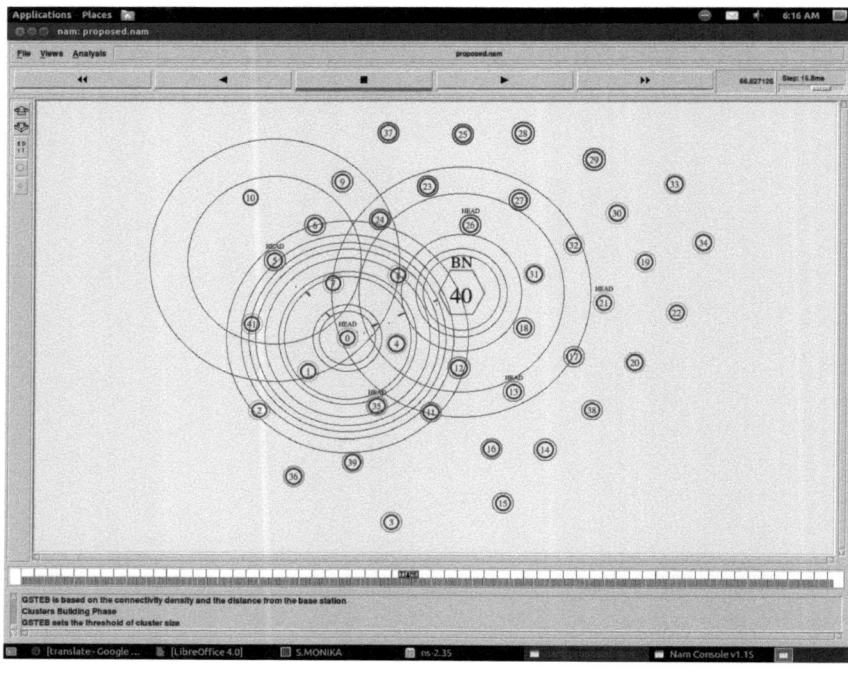

6. Multi Input and Output

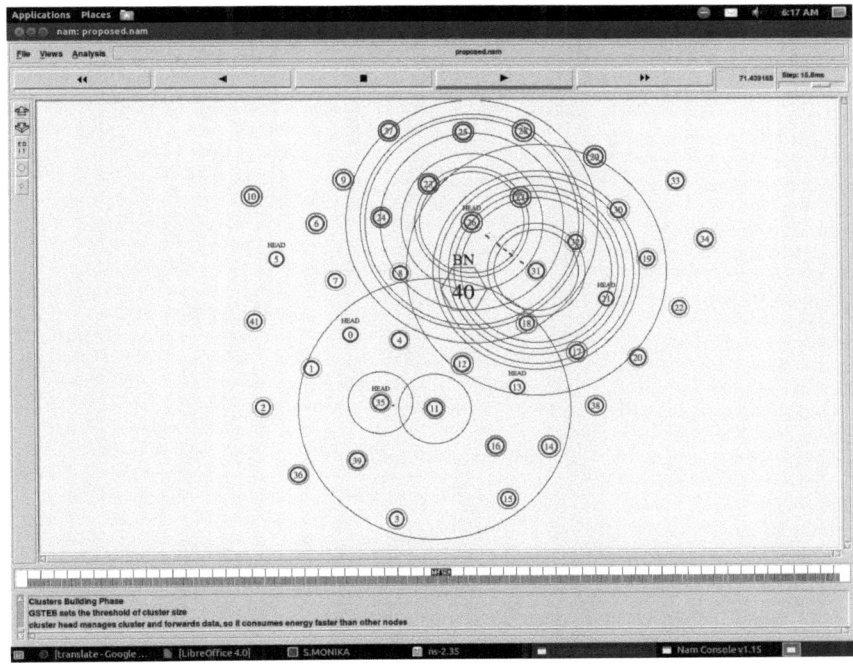

7. Master and Slave

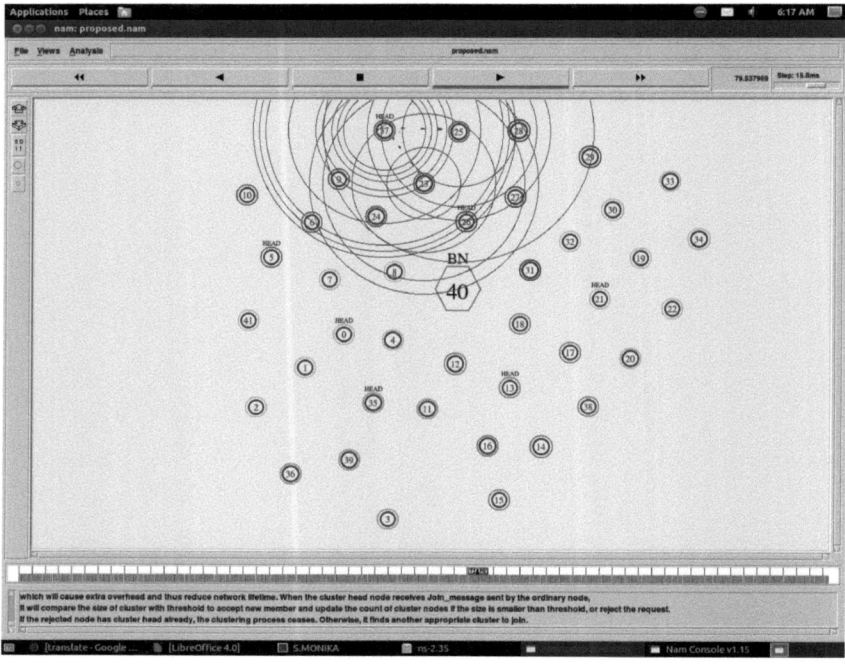

8. Re-electing Cluster Head

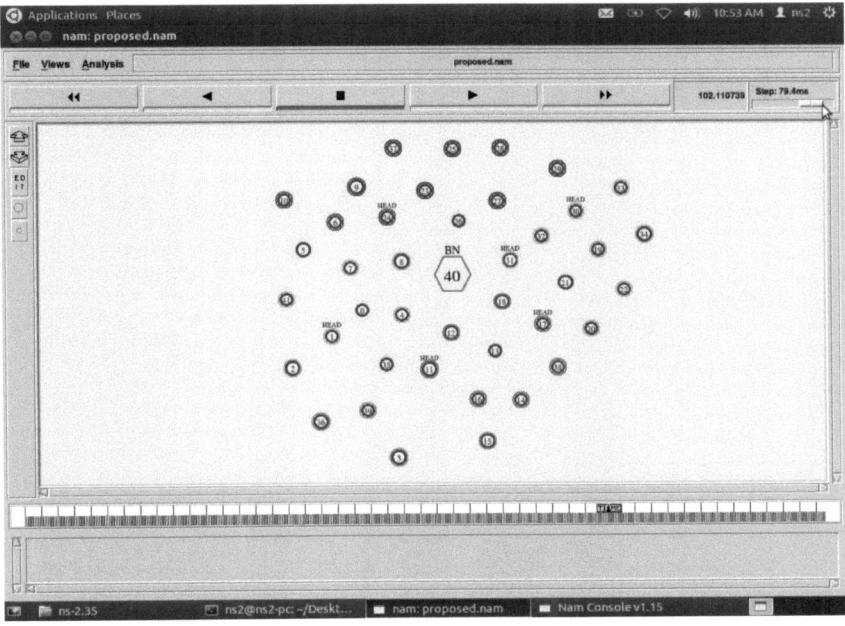

9. Re-election Cluster Head and Data Transmission

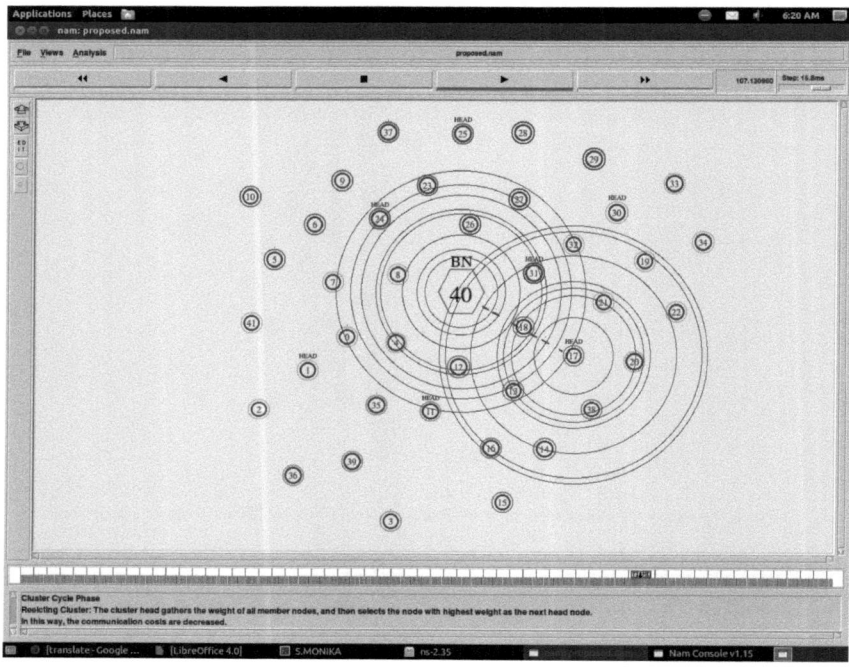

10. Re-electing Cluster Head and Tree Construction Data Transmission

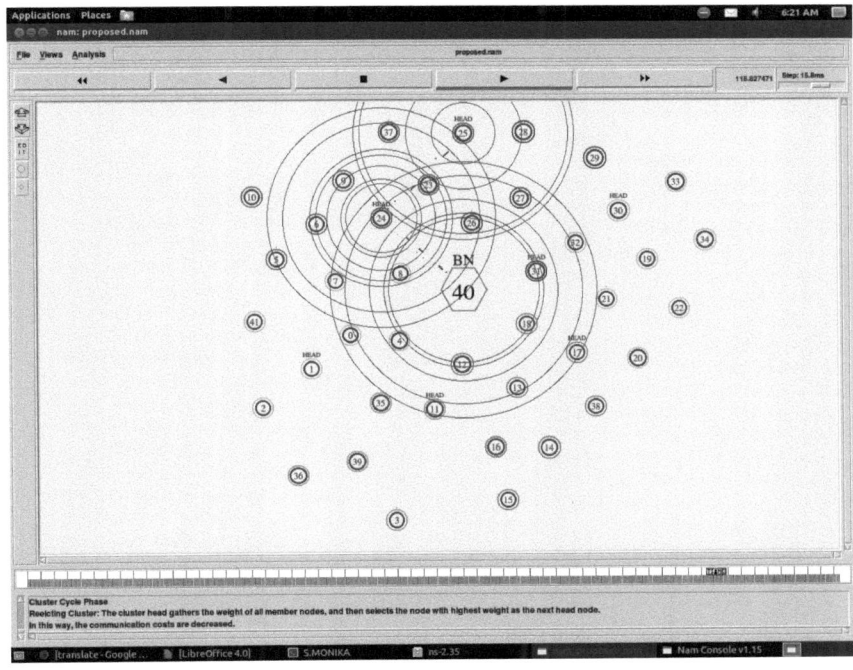

11. Selection of Source Node

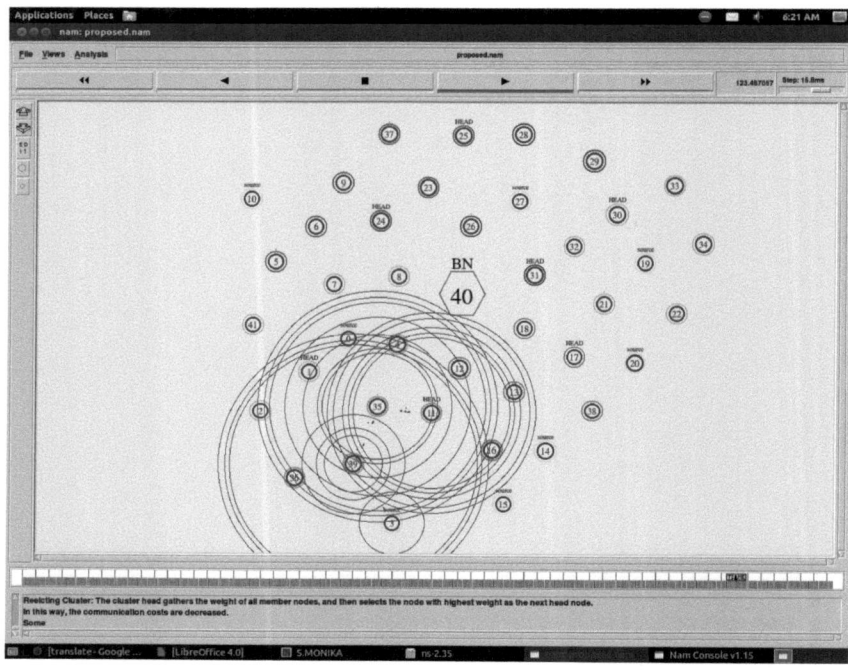

12. Next Round

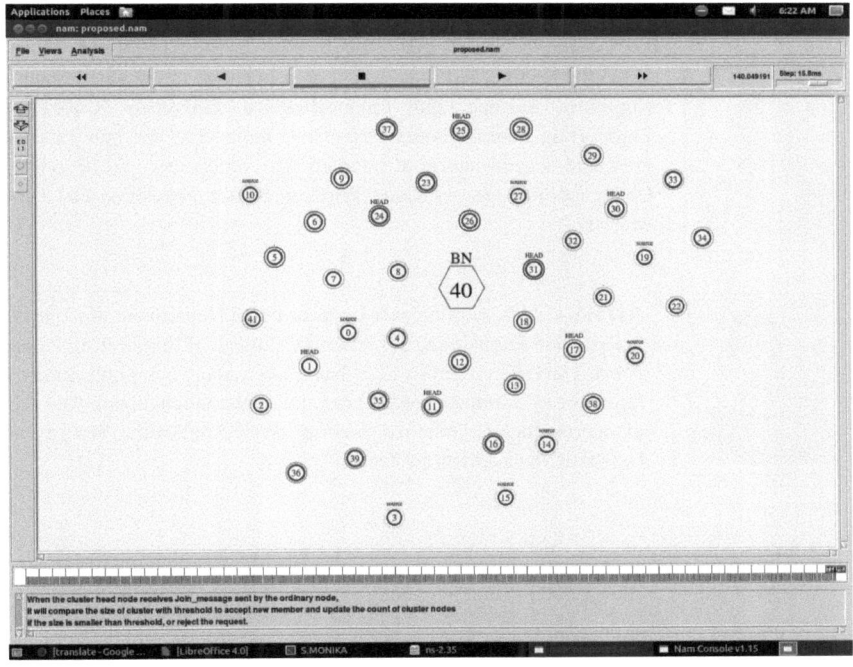

ABOUT THE AUTHORS

VAIYSHNAVI . M.P is working as a Lecturer in the Department of Computer Science and Engineering at University College of Engineering Panruti, Anna University, India. She has published and presented several papers at international conferences and in journals. Her areas of interest include Wireless Sensor Networks, and Cloud Security.

NITHYA .V is working as a Lecturer in the Department of Computer Science and Engineering at University College of Engineering Panruti, Anna University, India. She has more than five publications in International Journals, conferences and a International book. Her areas of interest include Wireless Security, Web Application Security and Database Management System.